Unit 3 — Using Scientific Skills

Published by Coordination Group Publications Ltd.

From original material by Richard Parsons.

Editors:
Ellen Bowness, Tom Cain, Gemma Hallam, Rose Parkin, Ami Snelling, Julie Wakeling.

Contributors:
Neil Atkin, Dr Giles R Greenway, Derek Harvey, Barbara Mascetti, Andy Rankin,
Philip Rushworth, Adrian Schmit, Pat Szczesniak, Sophie Watkins.

ISBN: 978 1 84146 759 7

With thanks to James Foster, Karen Hyde, Glenn Rogers and Caroline Russell for the proofreading.
With thanks to Katie Steele for the copyright research.

With thanks to Science Photo Library for permission to reproduce the photographs used
on pages 8, 14, 36, 51, 53, 59 and 71.

Groovy website: www.cgpbooks.co.uk

Printed by Elanders Ltd, Newcastle upon Tyne.
Jolly bits of clipart from Corell

Organisations that Use Science

Science contributes to your everyday life in loads of ways, most of which you probably haven't even thought about — the shampoo you use, the plastic used to make your lunchbox, and even the water you drink are all there because of science organisations...

Organisations that Use Science Can Benefit Society

There are loads of different science organisations and businesses — some make products and some provide a service. Here are some examples of how those products or services benefit society (i.e. you):

By Making Useful Products...

1) DRUGS

Pharmaceutical companies develop, test and manufacture drugs — drugs help us to fight off disease.

2) CHEMICALS

Chemical companies develop and produce things like fertilisers and paint — without these food would be more expensive (and art lessons would be a lot less interesting).

3) FOOD

Food manufacturers grow and process food (unsurprisingly) — they produce large amounts of safe food.

...And Providing Useful Services

1) KEEPING US HEALTHY

The health service includes hospitals, doctors, dental surgeries and pharmacists.

3) ANALYSING CHEMICALS

Lots of organisations have laboratories that do things like make sure the water is safe to drink and make sure products are of a consistent quality.

2) GLOBAL COMMUNICATION

Telecommunications companies make it possible to phone your mate in Australia.

4) EDUCATING PEOPLE

Schools, colleges and universities teach science (lucky for you).

5) PROVIDING ENERGY

Some companies generate and distribute energy — without electricity it'd be hard to do lots of things, let alone watch Neighbours.

These organisations also provide employment for millions of people and large companies generate lots of money for the country. Local organisations are just as important for the local economy.

Organisations may be Local, National or International

1) International organisations have sites in more than one country. They're usually big companies — some examples of those that use science are BP plc, Unilever plc and GlaxoSmithKline.

2) National organisations are based in just one country, and they distribute their goods or services throughout the country. In the UK there are organisations like the NHS and the Environment Agency, as well as nationwide shop chains. It's sometimes hard to tell whether a company is national or international, but you should be able to find this out from their website (if they have one).

3) Local organisations that use science include things like schools, colleges, health centres and dentists.

Science organisations — even better than sliced bread...

So there you have it — without a supply of people with science training and skills there would be no science organisations. That wouldn't leave us with very much. So, unless you want to live in a really rubbish future (with no cars, TV or even food), you'd better get learning — your country needs you.

Locating an Organisation

Ever wondered <u>why</u> an organisation is based <u>where</u> it is? Believe it or not, a lot of <u>thought</u> goes into where they're <u>located</u> and it's not just things like, "because it's round the corner from Aunty Flo's"...

There are <u>Lots</u> of <u>Factors to Consider</u>

Not all of these factors are relevant for all organisations, but the <u>general</u> reasons behind location are:

Raw materials

The presence of <u>raw materials</u> required for the process.

EXAMPLE: Breweries are often located next to supplies of <u>pure spring water</u> that is essential for the production of good quality beers and spirits.

Workforce

The availability of a <u>workforce</u> with the <u>right skills</u>.

EXAMPLE: Many high-tech companies (such as biotechnology companies) are based on <u>University science parks</u> because it's easy to recruit employees from the University.

Land

The <u>cost</u> of <u>land</u>.

EXAMPLE: The cost of land in the South East is more <u>expensive</u> than elsewhere — many companies have <u>relocated</u> to the Midlands and up North.

Energy

The availability of an <u>energy supply</u>.

EXAMPLE: Aluminium production is sited in Conwy (in Wales) because <u>hydroelectric power</u> can be readily produced there.

Transport links

Good <u>transport links</u> for delivery of raw materials.
EXAMPLE: Oil refineries are located around <u>ports</u> for supplies of crude oil from <u>tankers</u>.

Market

A <u>market</u> for the <u>product</u> or <u>service</u>.
EXAMPLE: Companies who manufacture dyes tend to be found in areas of textile manufacture.

Grants

Availability of <u>Government</u> or <u>European grants</u> to reduce the <u>start-up costs</u>.

There Could be <u>Effects</u> on the Local Environment

The previous page described some of the ways society benefits from organisations that use science. But because of the type of work they do, some organisations can have a <u>damaging effect</u> on the <u>environment</u>.

1) <u>Toxic pollution</u> — nasty chemicals, e.g. from a chemical works, could contaminate the environment.

2) <u>Visual pollution</u> — some factories and company <u>buildings</u> can be pretty <u>unsightly</u>. Also, things like chemical works and oil refineries are sometimes <u>illuminated</u> at night and are a source of <u>light pollution</u>.

3) <u>Noise pollution</u> — big trucks and big machinery are usually noisy. This can be a big problem for locals if the business operates 24 hours a day.

4) <u>Traffic congestion</u> — large businesses, e.g. a brewery, may need <u>frequent deliveries</u> of raw materials and <u>collection</u> of products — if this is done by road the lorries might cause <u>traffic congestion</u> in the local area and <u>damage</u> to road surfaces.

So you wouldn't put a tidal power station in the Sahara...

You may well be wondering <u>why</u> you need to know all this, well — soon you'll have to produce a <u>report</u> into an organisation, explaining the <u>reasons for its location</u> and its <u>effects on the local environment</u>.

Roles of Scientists

Of the UK's workforce, a massive four million people carry out jobs that use science.
You might be surprised how many different jobs there are in science...

Science Qualifications Offer a Range of Different Careers

This page covers just some of the areas in which scientific skills can be used. There are loads of
employment opportunities for people with scientific skills. There just isn't the space to list them all.

HEALTHCARE — e.g. doctors, dentists, nurses, pharmacists, radiographers, physiotherapists. There are also people who support these roles, e.g. medical physicists and lab technicians.

EDUCATION — e.g. secondary school science teachers, university and college lecturers.

ENGINEERING (the development of materials and technology) — e.g. chemical engineers develop paints and dyes, mechanical engineers develop machines.

PHARMACEUTICALS — e.g. research scientists develop, make and test drugs.

MANUFACTURING — e.g. analytical scientists are involved in quality control (making sure manufactured goods are of a consistent quality).

FOOD INDUSTRY — e.g. food scientists develop foods for supermarkets and food manufacturers, microbiologists test food to make sure it's safe.

SCIENCE QUALIFICATION

AGRICULTURE — e.g. research scientists look at new ways to produce foods or monitor standards of production, vets keep animals healthy.

There are loads of others — some scientists work for the police as forensic scientists, there are science editors, scientific patent lawyers, technicians who support the work of other scientists, goat breeders, orangutan urine collectors, and many, many more.

There are Major, Significant or Small Users of Science

1) Major users of science are people who use scientific skills as a large part of their job. They generally have a science-based qualification (see next page), e.g. science teachers, laboratory and research scientists, doctors and nurses.

2) Significant users of science are people who use scientific skills as part of their jobs — their training will have involved learning a fair amount of scientific knowledge, e.g. science editors and many of the healthcare professionals (such as dieticians).

3) Small users of science are people who use basic scientific skills as part of their jobs. They don't work in a science-based job and their training probably wouldn't have included that much science. Small users include hairdressers (carrying out allergy tests before applying hair dyes and bleaches), photographers (using chemical solutions to develop photographs), plumbers and electricians.

This is really just a guide though. There are no hard-and-fast rules about putting people into categories — different people might have different ideas about where they belong.

Rolls of scientists — how does that work then?

Hopefully now you can see that the possibilities really are endless. You could become an engineer, or even find yourself working as an editor for a company that makes revision guides, sharing your wealth of scientific knowledge with the young scientists of tomorrow. Still, don't expect that to make your mam happy — she'll still want you to join the navy.

Skills and Qualifications

Science is a <u>compulsory</u> subject in UK schools until age 16 — after that you can do <u>what you want</u>. Some (usually crazy) people decide that they haven't quite had enough and do even more science after the age of 16. There are loads of different options out there — <u>apprenticeships</u>, <u>degrees</u>, <u>NVQs</u>, the list goes on...

People Who Use Science Usually Have Special Qualifications...

1) People who are <u>major</u> scientific users will usually have a <u>degree</u> in science. This could be a <u>general degree</u>, e.g. in biology or chemistry, or a <u>specialised degree</u>, for example in forensic science or food science.

2) Some scientists, particularly those working in <u>research</u>, will have a <u>higher degree</u> — this can be either a masters (e.g. an MSc) or a PhD (scientists who have done a PhD are then called doctors).

3) For many careers you have to obtain special <u>professional qualifications</u>. Teachers have to do a <u>PGCE</u> (Postgraduate Certificate of Education), which is a special teaching qualification. People working in the <u>healthcare sector</u> have qualifications that test their understanding of science. They also may have to be '<u>registered</u>' with a <u>supervisory body</u> in order to practise.

4) Many organisations run their own <u>training schemes</u> (often linked to <u>Modern Apprenticeships</u> or <u>NVQs</u>) for careers like technicians and laboratory assistants.

5) Not all people who use scientific skills in their work will need to have science qualifications. This is the case if science skills only form a <u>small part</u> of their job (small users — see previous page).

... And a Wide Range of Skills

On top of <u>formal qualifications</u>, everybody who uses science in their work needs other <u>skills</u> — the exact skills required will depend on the <u>nature</u> of the job, but they might include things like:

1) <u>Research skills</u> — finding <u>information</u> from books, scientific journals or the internet.

2) <u>Communication skills</u> — getting your ideas across in a <u>clear</u> way.

3) <u>Numeracy skills</u> — being able to take <u>measurements</u>, carry out <u>calculations</u> and <u>analyse</u> data using <u>statistics</u>.

4) <u>IT skills</u> — using computer packages, e.g. to make <u>spreadsheets</u> and <u>databases</u>.

5) <u>Planning skills</u> — planning <u>investigations</u> that will be <u>successful</u> and hopefully give <u>good results</u>.

6) <u>Analytical skills</u> — breaking problems down into <u>smaller</u>, easier-to-solve chunks.

7) <u>Observational skills</u> — making <u>accurate</u> and <u>useful</u> observations of experiments and accurately <u>recording</u> results.

8) <u>Applying specialist knowledge</u> — <u>assessing</u> results and drawing <u>conclusions</u>.

9) <u>Team working skills</u> — working as a team is important for a lot of scientific work. Good team working skills will mean that the task can be completed to a <u>high standard</u> in an <u>efficient way</u>.

A typical team of scientists at work — remember, there's no 'I' in 'team'.

Next time you break one of your mam's vases...

...just tell her you were practising your analytical skills. You'll probably need skills like this no matter what kind of <u>job</u> you go for. But you're probably wondering just <u>why</u> you need to know about all this — well, one of the things you have to write about in your <u>report</u> is the <u>skills</u> and <u>qualifications</u> of scientists, so it's going to come in mighty <u>handy</u> for that. Also I thought that maybe you'd just like to know.

Report: Science in the Workplace 1

Well, now the section's over it's time to crack on with that <u>report</u> I've been blabbering on about.

You Need to Write a Report on Science in the Workplace

This will be the FIRST OF TWO reports that make up your portfolio for <u>UNIT 1: SCIENCE IN THE WORKPLACE</u>.

Your report will have <u>two bits</u> to it:

1) A <u>DESCRIPTION</u> of a <u>minimum</u> of <u>three</u> organisations that use science or scientific skills (and a more <u>in-depth study</u> of one of them), including:
   - <u>General information</u> about the <u>organisation</u> — what <u>products</u> they make (or what <u>services</u> they provide), where they're <u>located</u>, and whether they're <u>local</u>, <u>national</u> or <u>international</u>.
   - The <u>jobs</u> of those employed, and what <u>qualifications</u> and <u>skills</u> they have.

2) You also need to write about the <u>TYPES OF CAREERS</u> that are available in science. (You could link this to your workplace report by identifying the careers in the organisations you studied.)

Choose Your Organisations Carefully

1) It's no good picking three organisations that are <u>all the same</u> (e.g. three international drug companies) — it'd be pretty dull for you and won't get you great marks. Try to pick organisations from <u>different areas of science</u> (e.g. healthcare, environment and engineering) and try to pick one <u>local</u>, one <u>national</u> and one <u>international</u> organisation.

2) There are plenty of places you can look for <u>inspiration</u> — e.g. the phone book, the internet, newspapers (local and national), job adverts and the local job centre. Your friends and family might have some good ideas (or might even work for an organisation that uses science).

Find Out General Information from Websites

1) If they're a biggish company they'll probably have a <u>website</u>. This should tell you loads of the things you need to know, e.g. <u>what they do</u> and <u>where they're located</u> (and if it's in more than one country you know they're international).

2) If you can't find information about your chosen organisation on the internet you might have to <u>write</u> and ask for it. You could also prepare a <u>questionnaire</u> and send it to the organisation.

3) You'll get better marks if you <u>describe</u> things, rather than just stating them — don't just say, 'They make drugs' — instead describe what type of drugs they make and what they're used for etc.

For top marks you also need to:
- <u>explain why</u> the organisation is <u>located where it is</u>,
- <u>explain</u> its <u>importance</u> to <u>society</u>,
- <u>describe</u> how it <u>affects</u> the <u>local environment</u>.

Then Find Out About Science Careers

There are plenty of ways to find out about science careers, e.g. <u>job / career websites</u>, <u>company websites</u> (under the job / career link), your <u>local job centre</u>. It's worth looking out for <u>job descriptions</u> — these tell you what qualifications and skills you have to have to do that job. Some big employers and science institutions have dedicated careers websites, e.g. the NHS have www.nhscareers.nhs.uk.

Writing reports? — I thought this was Science not English...

There's loads of info out there — the hardest part is knowing <u>where to start</u>. If you're really struggling to find anything about a particular organisation early on then it might be better to pick a different one.

Avoiding Hazards

Scientific work can be <u>dangerous</u>. You need to be able to work <u>safely</u> in order to <u>prevent</u> accidents from happening. This applies to all workplaces, e.g. school and industrial labs.

There are Six Main Types of Hazard You Should be Aware Of

Hazards need to be <u>identified</u> so that they can be <u>avoided</u>.
So, first things first — what types of hazard are there?

Hmm... Where did my bacteria sample go?

1) <u>MICROORGANISMS</u> — these are a particular problem in <u>microbiology labs</u>. The biggest hazard is coming into contact with microorganisms that can <u>cause disease</u>, e.g. <u>viruses</u> and <u>bacteria</u> (see page 25 for more).

2) <u>RADIATION</u> — this is emitted by <u>radioactive materials</u>. The effect on body tissues can be devastating (<u>nausea</u>, <u>weakened immune system</u>, even <u>death</u>) so you need to take precautions (see p.8). This <u>hazard symbol</u> is used to label a radioactive source.

3) <u>CHEMICALS</u> — There are different types of hazardous chemical, each with a <u>hazard warning symbol</u>:

 <u>Oxidising</u> — These provide <u>oxygen</u>, which allows other materials to <u>burn more fiercely</u>, e.g. liquid oxygen.

 <u>Toxic</u> — Can cause <u>death</u> either by being <u>swallowed</u>, <u>breathed</u> in, or <u>absorbed</u> through the skin, e.g. cyanide.

 <u>Flammable</u> — <u>Catch fire</u> easily, e.g. petrol.

 <u>Corrosive</u> — <u>Attacks and destroys materials</u>, particularly <u>living tissues</u> such as eyes and skin, e.g. sulfuric acid.

 <u>Irritant</u> — Not corrosive but can cause <u>reddening or blistering</u> of the skin, e.g. bleach.

4) <u>ELECTRICITY</u> — electrical hazards include long or frayed cables, cables touching something hot or wet, damaged plugs, overloaded sockets and machines without covers. Electrical hazards can cause <u>electric shocks</u> that may lead to <u>burns</u> or <u>death</u>.

5) <u>GAS</u> — it's important to make sure all gas hoses and taps are in <u>proper working order</u>. Gas is <u>flammable</u>, but thanks to its <u>smell</u> it's usually pretty obvious if someone leaves a tap on. However, left unnoticed it can cause <u>suffocation</u> or an <u>explosion</u>.

6) <u>FIRE</u> — many things in the workplace can cause fires, particularly in the petrochemical industry. <u>Damaged electrical appliances</u> are a big culprit. For more on fire see page 10.

Safety Signs Warn You of Hazards in the Workplace

<u>Safety signs</u> give health and safety information in the normal course of work.
There are <u>four colours</u> of safety sign — they have specific meanings.

 <u>Blue</u> — <u>mandatory</u> sign. Instruction <u>must</u> be followed.

 <u>Yellow</u> — <u>warning</u> sign. Take <u>care</u>.

 <u>Green</u> — <u>safety</u> information. E.g. fire exit / first aid point.

 <u>Red</u> — <u>prohibition</u> sign. Action shown must <u>not</u> be carried out.

Risk — not just a thrilling board game...

You'll need this stuff for doing your <u>report</u>, and if you don't follow my advice, then you're just asking for <u>trouble</u>. Don't come crying if you catch <u>hepatitis</u> or <u>electrocute</u> yourself on an overloaded socket.

Avoiding Hazards

Once you've identified your hazards, the next step is to prevent accidents from happening.

The Risk of Injury can be Reduced in Four Ways

The majority of accidents happen because of human error.
The chances of an accident happening are reduced by things like:

1) Proper behaviour — This includes things like not running around the laboratory, and holding scissors or blades in the correct way. It's also important not to eat, drink or smoke in the lab — this is especially important when handling microorganisms and toxic or flammable chemicals.

2) Using equipment properly — all equipment will have instructions and it's important to follow these exactly. Improper use might damage the equipment, but could also lead to serious injury.

3) Using protective and safety equipment — if it's needed, protective equipment has to be provided — it's the law (see below). Lab coats protect your clothes and safety glasses prevent chemicals or flying glass from damaging your eyes. In some scientific workplaces such as hospital laboratories it's important for workers to wear masks and gloves to prevent them being infected with nasty diseases.

4) Following correct procedures — when carrying out an experiment you should always have a well-planned procedure before beginning. It's important to follow the procedures, e.g. using too much of a substance could result in injury if the substance is toxic or flammable.

When carrying out any experiment it's important to have the right protective equipment.

Workplaces are Governed by Health and Safety Regulations

Health and safety legislation is there to provide employees with a safe and healthy working environment. There's a long list of regulations, which cover things like:

1) General health and safety — the workplace must not be a big risk to the people who work there.

2) Electricity — covering the safe use of electricity (obvious really).

3) Personal protective equipment — employers have to provide protective equipment free of charge where it's needed, e.g. goggles, gloves, helmets etc.

4) Control of hazardous substances — employers are required to label all hazardous substances. They must also have a policy on how the risks of using them can be kept as low as possible.

Protective equipment like helmets and safe footwear must be provided.

> Employers are legally required to assess the risks within the workplace. A risk assessment is an examination of what could cause harm in the workplace. There are five stages to a risk assessment:
> 1) Look for hazards.
> 2) Assess who may be harmed and how.
> 3) Decide what action, if any, needs to be taken to reduce the risk.
> 4) Document the findings.
> 5) Review the risk assessment regularly.

Health and safety officials can enter workplaces at any time to carry out an inspection. Where they find problems they can issue instructions for improvements or stop work being carried out altogether. If breaches of health and safety regulations are really serious employers could end up in court.

It's better to be safe than sorry...

All this might seem pretty dull but you'll be thanking your lucky stars when some rather fetching safety glasses save you from losing your eye in a freak beaker accident.

Radioactive Substances

Many industries use radioactive materials, but they can be <u>dangerous</u> because they <u>produce radiation</u>. This can <u>damage</u> your body, so it's pretty important to know how to <u>handle</u> and <u>dispose</u> of these materials in a safe way.

Care Must be Taken When Handling Radioactive Substances

On average, there's one serious incident (resulting in <u>death</u> or <u>serious injury</u>) involving <u>radioactive material</u> in the world each year. The figures are so low because laboratories take <u>precautions</u> when handling radioactive material.

1) <u>Never</u> allow <u>skin contact</u> with a source — always handle with <u>tongs</u>.
2) Keep the source at <u>arm's length</u> to keep it <u>as far</u> from the body <u>as possible</u>.
3) Keep the source <u>pointing away</u> from the body and <u>avoid looking directly at it</u>. (And <u>don't</u> point it at <u>anyone else</u> whilst you're doing this.)
4) <u>Always</u> keep the source in a <u>lead-lined box</u> and put it back in <u>as soon</u> as the experiment is <u>over</u>.

Extra Precautions are Needed for Industrial Workers

1) Radioactive substances are used widely in <u>industry</u> — <u>hospitals</u> use them for things like <u>radiotherapy</u> and <u>diagnostic imaging</u>. They're also used for <u>sterilising food</u> and in <u>nuclear power stations</u>.
2) In hospitals workers like <u>radiographers</u> have to keep their <u>radiation exposure</u> to a <u>minimum</u>.
   - They <u>leave the room</u> or stand behind a <u>protective screen</u> when doing things like carrying out radiotherapy (which uses a type of radiation).

3) Some <u>nuclear power station workers</u> also have to take extra precautions:
   - To prevent workers being exposed to radiation <u>lead-lined suits</u>, <u>lead or concrete barriers</u> and <u>thick lead screens</u> are often used.
   - Some workers have to wear <u>full protective suits</u> to prevent <u>radioactive particles</u> from being <u>inhaled</u> or getting trapped <u>under their fingernails</u> etc.
   - Working in highly radioactive areas is often <u>too dangerous</u> for people so workers use <u>remote-controlled robot arms</u> to move things about. Cool.

STEVE ALLEN / SCIENCE PHOTO LIBRARY

Radioactive Material Used in Schools Produces Low-Level Waste

1) The majority of radioactive waste produced by hospitals, universities, schools and colleges is <u>very low-level waste</u>. It usually includes things like gloves, masks, bench coverings, paper towels etc. (Nuclear power stations produce <u>high-level</u> radioactive waste, which has to be dealt with in a special way.)
2) Very low-level waste can be sealed in a <u>strong plastic bag</u>, then <u>thrown away</u>. It <u>is</u> safe to dispose of this waste in landfill sites because the level of radiation is so <u>low</u>.
3) Disposal of radioactive <u>sources</u> (the material that's actually used for experiments) is slightly different. Solid sources (like the ones you'll use at school) should be put in a <u>small container</u> and filled up with <u>plaster of Paris</u>. The container <u>should not</u> be labelled to show that it contains a radioactive source (if you labelled it some nosey parker might pick it up). The container can then be put in with normal rubbish (but you can't dispose of sources this way more than once a week).

Robot arms — my preferred dancing style...

It might sound a bit careless just to put radioactive waste out for the bin men but it's only <u>very</u> low-level waste. <u>Nuclear power plants</u> have to be extra <u>careful</u> with their waste. They can't just go around putting weapons grade plutonium in plastic bags and chucking them in the bin — that'd just be wrong.

First Aid

Now, if you've been <u>paying attention</u> over the last few pages then hopefully you'll <u>never need</u> to use what you learn on this page. Having said that, there's always going to be some <u>idiot</u> clowning around, <u>causing trouble</u> for the rest of us — so it's probably for the best if you have a <u>good read</u> over this stuff anyway.

First Aiders Could be the Difference Between Life and Death

It's a good idea, but <u>not a legal requirement</u>, to have as many people as possible trained in basic <u>first aid</u>. They could be vital in saving the life of a person in <u>any situation</u>, e.g. at work, in the street or at home.

Training courses in basic first aid are provided by <u>St John Ambulance</u>, <u>St Andrew's Ambulance Association</u>, and the <u>British Red Cross</u>. All these organisations have websites and can be found in the phone book.

In Any First Aid Situation Follow a Clear Plan of Action

This will stop you placing <u>yourself</u> in danger and will help you to respond in the <u>right way</u>:

1) <u>Assess the situation</u> — what has happened? Is anyone still in danger?
2) <u>Make the area safe</u> — protect yourself and the casualty from danger.
3) <u>Give emergency aid</u> — give appropriate first aid (see below). If there's more than one casualty, the ones with <u>life-threatening</u> conditions should be treated <u>first</u>.
4) <u>Get help</u> — once the casualty has been <u>stabilised</u> call an <u>ambulance</u>.

You Need to Know the Treatment for Common Injuries

There are <u>seven</u> common injuries that you might encounter in the laboratory. You need to know <u>what they are</u>, <u>what basic first aid should be given</u> and when it would be <u>unsafe to give first aid</u>.

1) <u>Heat burns and scalds</u> — if they're <u>minor</u> flood the injured part with <u>cool water</u> for at least 10 minutes, then cover with a <u>sterile dressing</u>. If they're <u>pretty serious</u> cool, damp cloths should be used instead and you should ring an <u>ambulance</u>.
2) <u>Chemical burns</u> — flood the injured part with water for at least 20 minutes, remove any <u>contaminated clothing</u> and arrange for the casualty to be sent to <u>hospital</u>. You should <u>not</u> attempt to give first aid if there are <u>chemical fumes present</u> or if there has been significant <u>chemical spillage</u>.
3) <u>Poisoning due to fume inhalation</u> — the priority is to get the casualty into <u>fresh air</u> and to get <u>medical help</u>. You should not attempt to move the casualty if there are <u>fumes in the area</u>.
4) <u>Poisoning due to swallowing</u> — the casualty needs to go straight to <u>hospital</u>. Never attempt to make the casualty <u>vomit</u> and never give them anything to <u>drink</u> (although <u>small sips</u> of water are OK if they've swallowed something <u>corrosive</u>).
5) <u>Electric shock</u> — turn off the <u>electrical supply</u> before doing anything else (don't touch the victim until you've done this), then ring an <u>ambulance</u>. If the casualty stops breathing you need to be prepared to give <u>rescue breathing</u> (mouth-to-mouth resuscitation) and <u>chest compressions</u>.
6) <u>Cuts</u> — clean the wound under <u>running water</u>, raise the injured part if possible and apply a <u>dressing</u> (<u>don't</u> try to pull objects out of wounds — pad around them and bandage over the top, then send the casualty to <u>hospital</u>).
7) <u>Particles or chemicals in the eye</u> — particles or chemicals should be <u>flushed out</u> of the eye using lots of <u>sterile water</u>. You need to do this for 10 minutes, then bandage the eye before sending the casualty to <u>hospital</u>.

First aid — relief for someone in need of a brew...

An important thing to remember about first aid is that you shouldn't do anything that might put <u>you</u> or the <u>casualty</u> at <u>risk</u>. Never do anything that you're <u>not sure about</u> just for the <u>sake</u> of doing something.

Fire Prevention

Fires are responsible for many deaths every year, but only 6% of these occur in the workplace. The low figure is all thanks to things like <u>fire instructions</u>, <u>sprinklers</u> and <u>extinguishers</u>...

Fire Instructions <u>Tell You</u> <u>What To Do</u> <u>in the Event of a Fire</u>

Fire instruction notices should be displayed at <u>prominent</u> points in a building — they tell you the <u>quickest route</u> to leave the building and <u>where to assemble</u>. Make sure you're <u>familiar</u> with your school lab ones. Also make sure you know what the fire alarm <u>sounds like</u> — it should be tested at least <u>once a week</u>.

If the fire alarm <u>sounds</u> you should:

1) Leave the building by the <u>quickest escape route</u> — <u>never</u> use lifts, escalators or revolving doors.

2) Go to the designated <u>assembly point</u> and wait there until a fire warden takes a roll-call — <u>don't</u> wander off or go home, or somebody might <u>re-enter</u> a burning building to <u>look for you</u>.

If you <u>discover</u> a fire you should:

1) <u>Sound</u> the fire alarm (usually by smashing the glass at a fire alarm point).

2) <u>Call the fire brigade</u> (though some alarms will automatically alert the fire brigade).

3) If the fire is <u>small enough</u>, use a hand-held extinguisher to tackle it (the different types of extinguisher are listed below) — but <u>never</u> put yourself at <u>risk</u> in attempting to put a fire out, and <u>always</u> stand <u>between</u> the fire and your escape route.

4) Leave the building by the <u>quickest escape route</u> and report to the <u>assembly area</u>.

Fire Doors <u>and Sprinkler Systems</u> <u>can Stop Fire Spreading</u>

Fires can spread easily through <u>open areas</u> such as <u>corridors</u> and <u>stairwells</u>. There are <u>two</u> common features installed in the workplace to <u>slow down</u> the spread of fire.

1) <u>Fire doors</u> act as barriers to hold back <u>smoke</u> and <u>flames</u>. Fire doors must be kept <u>shut</u> at all times or be fitted with <u>automatic closing devices</u> — <u>never</u> wedge a fire door open.

2) <u>Sprinkler systems</u> are usually installed in <u>high-risk</u> areas, such as <u>storerooms</u>. They're very effective at <u>containing fires</u> — they spray water from the ceilings. But, they're <u>expensive</u> to install and maintain, they need a water supply at <u>high pressure</u>, and if there's a minor fire (which can be put out with a fire extinguisher) they can cause a lot of unnecessary <u>mess</u> and <u>damage</u> to equipment and stock.

Know Which Type of Fire Extinguisher to Use

There are <u>six types</u> of hand-held fire extinguisher. All new fire extinguishers are painted red with a <u>colour-coded</u> band or panel to identify its <u>contents</u> and the <u>type of fire</u> it can be used on. It's important to use the <u>right fire extinguisher</u> for the type of fire — you could make the <u>fire worse</u> if you use the <u>wrong one</u>.

Red — WATER
Used for: wood, paper, coal.

Black — CARBON DIOXIDE
Used for: wood, paper, coal, liquids, electrical equipment.

Green — VAPORISING LIQUID
Used for: liquids, electrical equipment.

You can also use a <u>fire blanket</u> to smother a fire if it's small and self-contained.

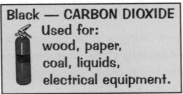

Cream — FOAM
Used for: wood, paper, coal, liquids.

Blue — DRY POWDER
Used for: wood, paper, coal, liquids, gases, electrical equipment.

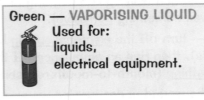

Yellow — WET CHEMICAL
Used for: cooking oil and fats.

Fire doors — bet they have hot handles...

Remember — it's dead important to use the <u>right kind</u> of extinguisher for a fire. You should <u>never</u> use <u>water</u> on an <u>oil fire</u> (e.g. a chip pan fire) — it could make the fire even <u>worse</u> and cause an <u>explosion</u>.

Report: Science in the Workplace 2

Now on to the unpleasant subject of <u>assessment</u> — those lovely examiners want you to write a report on working safely in the workplace. It's probably a good idea to base this report on the <u>same</u> organisation you picked for your first report.

You Need to Write a Report on Working Safely

This will be the SECOND OF TWO reports that make up your portfolio for <u>UNIT 1: SCIENCE IN THE WORKPLACE</u>.

Your report will have <u>three bits</u> to it:
1) A description of how <u>HAZARD ASSESSMENT</u> is managed for a <u>scientific workplace</u>.
2) Details about <u>FIRST AID</u> for that workplace.
3) Details about <u>FIRE PREVENTION</u> for that workplace.

Find Out How They Assess Hazards

1) To get <u>specific information</u> about health and safety in your chosen scientific workplace you'll need to write a short <u>questionnaire</u>, <u>phone them</u> or <u>visit in person</u>.

2) By law, firms with five or more employees must have a written <u>health and safety policy</u> — you could try asking (politely, of course) for a copy. It should contain a lot of the information you're after.

3) There are plenty of <u>questions</u> you could ask to get the information you need:
   - <u>Risk assessment</u> — who's responsible for making risk assessments? <u>How often</u> are risk assessments reviewed? (You could include a copy of a risk assessment in your report.)
   - <u>Machinery and equipment safety</u> — what's the procedure for checking electrical and mechanical equipment? Is there a maintenance logbook?
   - <u>Hazardous substances</u> — what substances are used? <u>Who's exposed</u> to these substances? What <u>storage</u> arrangements and <u>safety precautions</u> are in place?
   - <u>Protective clothing and equipment</u> — what items of protective clothing are provided? When should these items be used? (You should find this information in the risk assessments.)
   - <u>Safety information</u> — what information on health and safety is available to staff?
   - <u>Job training</u> — is there a safety induction programme?
   - <u>Safe working practices</u> — How do employers make sure health and safety procedures are followed?

Then Find Out About First Aid and Fire Safety

There are plenty of questions you could ask to get this information:
   - <u>First aid</u> — where are the <u>first aid boxes</u>? Where is the <u>accident book</u> kept? What is the policy on having trained <u>first-aiders</u>?
   - <u>Fire safety</u> — what are the <u>emergency evacuation procedures</u>? What <u>hand-held</u> fire-fighting equipment is available?

For top marks you also need to:
   - <u>research</u> working safely in your <u>school or college</u> (do all the things above again),
   - <u>compare</u> this to your chosen <u>workplace</u> (point out how they are <u>different</u> and how they are <u>similar</u>).

Reports, investigations? Who do you think you are? — Poirot?

There's no magic number of words that your report must be — it has to be long enough to get your <u>point across clearly</u>, but not so long that you fill it with <u>waffle</u>. Good luck.

A Balanced Diet

You'll have heard all about the need for a healthy, balanced diet time and time again. But read on and you'll find out exactly why we need to eat such a variety of food. Then maybe you'll start eating right too.

A Balanced Diet Supplies All Your Essential Nutrients

1) <u>Nutrients</u> are the chemical elements and compounds that our bodies need to keep <u>working</u>.
2) We need a <u>variety</u> of nutrients for functions like <u>respiration</u>, <u>movement</u>, <u>growth</u> and <u>repair</u> of tissues.
3) Having <u>too much</u> or <u>too little</u> of any nutrient can cause <u>health problems</u>.

> To get the <u>right amount</u> of nutrients, you need to eat a <u>balanced diet</u>, which means:
> * Eat <u>plenty</u> of starchy carbohydrates, fruits and vegetables.
> * Eat a <u>moderate</u> amount of dairy products, meats or other high protein foods, e.g. nuts.
> * Eat a <u>small</u> amount of fatty and sugary foods.

Here's What Nutrients Do and Where They're Found

Here's a little run-down on the five main <u>nutrients</u> you need in your diet:

Nutrients	Important functions in the body	Foods in which they're found
Carbohydrates (sugary and starchy foods)	Provide energy for all the functions of your body.	Starchy foods, e.g. bread, potatoes, cereal, pasta, rice Sugary foods, e.g. sweets, chocolate
Fats (saturated and unsaturated)	• Provide stored energy. • Insulate and help regulate body temperature. • Cushion vital organs against shock, e.g. kidneys. • Provide the fat-soluble vitamins — A, D, E and K.	Margarine, butter, oils, cream, fried foods, crisps, cakes, nuts, seeds
Proteins	• Provide energy. • Used to build new cells for growth, repair of body tissues and muscle development.	Meat, fish, cheese, eggs, pulses, nuts
Vitamins and minerals	Many different essential body processes. (Look at the next page for more info.)	Vegetables, fruits, dairy products
Water	Involved in all the chemical reactions of life.	Everything contains water.

Don't Forget About Fibre

1) <u>Fibre</u> isn't a nutrient, but it's something you need <u>plenty</u> of in your diet.
2) <u>Fibre</u> is a name for the bits of plants that <u>aren't</u> broken down by the digestive system when you eat them — they just pass <u>straight through</u> you.
3) Foods rich in fibre include <u>cereals</u>, <u>pulses</u>, <u>fruits</u> and <u>vegetables</u>.
4) We need lots of fibre to keep our <u>digestive system</u> in good working order — fibre helps prevent <u>constipation</u>, and maybe even helps prevent <u>bowel cancer</u>.

Health warning:
Too much bran causes the opposite of constipation.

A balanced diet — 1 kg of sweets for every 1 kg of chips...

A little reminder for you... You need to eat carbohydrates, saturated and unsaturated fats, proteins, vitamins, minerals, water and fibre every day. You also need to be careful about the <u>amounts</u> you eat of each. Deciding what to eat is a full-time job — goodness knows how we have time to actually eat.

Vitamins and Minerals

I bet you were wondering what those mysterious vitamins and minerals do... You weren't wondering? Oh.

We Need Vitamins to Stay Healthy

Different vitamins have different jobs in the body. These are the most important ones to learn about:

Vitamin	Important functions in the body	Foods in which it's found
A	• Maintains healthy eyesight. • Keeps mucous membranes (in the ears and nostrils) free from infection.	Milk, eggs, carrots, spinach
B vitamins	• Release energy from carbohydrates in foods. • Maintain the nervous system.	Potatoes, bananas, cereal, lentils, yeast
D	• Maintains healthy teeth and bones. • Aids the absorption of calcium and phosphorus.	Fish, egg yolks, fortified cereal and dairy products (have vit D added to them)
K	Helps blood clot.	Spinach, broccoli, cabbage
C	• Maintains the immune system. • Aids the absorption of iron. • Maintains the skin, and the linings of the digestive system.	Citrus fruits, potatoes, tomatoes

Vitamin D is also produced by the body when exposed to sunlight.

We Also Need Minerals

Like vitamins, different minerals also have different jobs in the body. Learn about these ones:

Mineral	Important functions in the body	Foods in which it's found
Iron	Helps the body to make haemoglobin (found in red blood cells) which is needed for transporting oxygen around the body.	Meat, fish, green vegetables, dried fruit
Calcium	Maintains healthy teeth and bones.	Milk, cheese, yoghurt, green leafy vegetables
Phosphorus	Helps release energy from food.	Dairy products, meat, fish
Zinc	Needed to make many enzymes work and for wounds to heal. (Enzymes are biological catalysts that make all the reactions in the body work properly.)	Red meat, shellfish, seeds, green vegetables

Vitamin B group — sounds like a dodgy band from the 70s...

Remember — if you're a vegetarian, you've got to find ways of getting the protein, zinc and iron that you're missing out on by not eating meat. That means eating more things like green veggies, pulses and dairy products. If you're worried, you could take dietary supplements, but lovely fresh food is best.

Diet Problems

OK, so you've got to eat all those vitamins, minerals and other goodies. But you'd probably rather not, I bet. Well, here's one excellent reason why you should eat a proper balanced diet — bleeding gums...

Not Eating Enough Vitamins Can Make You Ill

If you don't eat enough of certain foods, you can get vitamin deficiencies. For example, sailors used to get scurvy, a vitamin C deficiency, because they didn't have any fresh fruits to eat on long sea journeys.

You need to learn these examples of vitamin deficiencies:

Vitamin	Symptoms of deficiency
A	• Problems adjusting eyes to dim light. • Dry skin and mucous membranes.
B vitamins	• Anaemia. • Mouth sores. • Nerve cell degeneration.
C	• Bleeding gums. • Poor healing of cuts and wounds. • Weakening of blood vessels.
D	Weak teeth and bones, which may deform due to body weight.

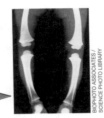

BIOPHOTO ASSOCIATES / SCIENCE PHOTO LIBRARY

Too Much of Some Nutrients Can Cause Problems Too

Even if you feel perfectly healthy now, your diet could be causing you problems for later life. These three are the biggies you should try to not eat too much of:

SATURATED FATS

Saturated fats raise cholesterol in the blood. Cholesterol is a fatty substance that's essential for good health. It's found in every cell in the body. However, a high level of it in the blood causes an increased risk of problems like coronary heart disease. Saturated fats should be eaten in moderation.

SALT

Eating too much salt can cause high blood pressure (for some people). High blood pressure increases the risk of heart disease. It's hard to keep track of how much salt you eat — most of it is probably in processed foods (such as breakfast cereals, soups, ready meals etc.). On food labels, it's usually listed as sodium.

SUGAR

Sugar provides energy, and that's all. If you exercise and use up all the energy provided by sugar, that's fine. If you don't, any leftover energy from sugar is stored by the body as fat. A high sugar diet could lead to raised cholesterol in the blood, and obesity, which can both cause heart disease.

So... too much of these nutrients could cause heart disease and obesity in later life. Obesity in turn could cause type 2 diabetes — a condition where the body can't absorb glucose properly. It has to be controlled with a special diet.

The best advice in the world — eat more fruit and veg...

If reading this page has made you panic and develop all sorts of symptoms in a fit of hypochondria, then don't worry. That's normal. When writing about vitamin B, I discovered a possible mouth ulcer and panicked, then when writing about vitamin D, I felt a bit weak, and panicked. Anyway, erm, learn it all.

Lifestyle and Diet

Here's how to change your life, get healthy and live longer. Dump that fast food and get on your bike...

Everyone Needs _Different_ Amounts of Energy

1)   We get the <u>energy</u> we need from the <u>food</u> we eat.
2)   Energy in food is often measured in <u>kilojoules</u> (kJ), or in <u>Calories</u> (kcal).
3)   The more <u>active</u> you are, the more <u>energy</u> you need, as you can see from the table below.

	Pensioner	Office worker	Cycle courier	Footballer	Labourer
Typical energy needed per day (kcal)	2000	2500	3500	4000	4500

4)   If you eat <u>more</u> calories than you <u>use</u> in activity, what's left over is <u>stored as fat</u> and you gain weight.
5)   <u>Calorie-controlled diets</u> involve eating <u>fewer</u> calories than you use in activity, to use up your <u>stored fat</u> and lose weight. These diets also suggest <u>cutting out</u> fatty and sugary foods that are high in calories.

Get Your Energy from _Healthy, Nutritious Foods_

1)   We could get all the energy we need by only eating <u>chocolate cake</u> — but we <u>wouldn't</u> be healthy.
2)   Chocolate cake has a <u>low nutritional value</u> — it lacks vitamins, minerals and nutrients like protein.
3)   We should get our energy from foods with a <u>good nutritional value</u>.
4)   Health professionals usually recommend that <u>more than half</u> of our calories should come from <u>carbohydrates</u>, <u>less than a third</u> should come from <u>fats</u>, and the rest should come from <u>protein</u>.

Fast Food Companies _Have Been Blamed for Making Us Fat_

<u>Fast food</u> is prepared and served <u>quickly</u> at fast food restaurants. Burgers, hot dogs, chips and takeaways are typical examples of fast food.

Most fast food has a <u>low nutritional value</u>, and is highly <u>processed</u>, with very high levels of <u>fat</u>, <u>sugar</u> and <u>salt</u>.

This fatty, greasy burger is probably 80% cardboard anyway.

Some fast food companies have been getting into <u>trouble</u> for marketing unhealthy products at <u>children</u>:

1)   '<u>Marketing</u>' means <u>promoting</u> products, especially by <u>advertising</u>, to try and <u>influence</u> us to buy things.
2)   Fast food companies have been <u>criticised</u> for marketing their products to <u>children</u>, for example by using <u>cartoon characters</u> in promotions, and giving away <u>free toys</u> with food.
3)   A person's food <u>preferences</u> are formed during childhood, so marketing fast food at children could have <u>long-term effects</u> on their eating habits and health.

Do this

4)   However, we all have a <u>choice</u> about what we eat, so maybe we <u>shouldn't</u> <u>blame</u> fast food companies. Instead, we can avoid becoming overweight by making healthy <u>lifestyle choices</u> — exercising regularly, eating healthily, and limiting our intake of fast foods.

Eat this

OK clown, drop the kiddie meal and put your hands up...

Did you notice when all the big fast food companies suddenly started selling <u>salads</u> and <u>carrot sticks</u> instead of just burgers and fries? That was just after they got accused of making the children of the world fat. I can't say I've ever been to a fast food company to have a salad, but hey, it's a nice gesture.

Food Additives

It's amazing what food companies can do with additives. They can take a perfectly normal piece of potato, for example, and make it <u>taste</u> like chicken, <u>look</u> like a tomato and <u>last</u> for years and years. Brilliant.

There are Loads of Different Additives that Do Different Things

<u>Processed</u> foods (e.g. soups, ready meals etc.) tend to contain <u>more</u> additives than <u>fresh</u> foods. Here's <u>why</u> additives are put into those foods:

- to improve the <u>taste</u>
- to improve the <u>appearance</u>
- to increase the <u>shelf-life</u>

Here are the six main types of <u>additives</u> you need to know about:

Type of additives	What they do	Examples	Foods they're in
Antioxidants	Preserve oils and fats in food.	Vitamin C	Cakes, margarine
Flavourings and flavour enhancers	Give food a certain taste or smell.	Monosodium glutamate (MSG)	Soft drinks, crisps, soups, sauces
Colourings	Give food a certain colour.	Tartrazine	Fruit squash, fizzy drinks, sweets
Preservatives	Prevent or slow down food spoilage processes.	Benzoic acid	Meat products, bread, beer
Sweeteners	Add a sweet flavour without using sugar.	Aspartame	Soft drinks, low-calorie foods
Thickeners	Increase viscosity (thickness) of foods.	Starch	Jam, ice cream, sauces

Additives Have to be Tested to Make Sure They're Safe

Additives used in foods have to be <u>tested for safety</u>:

1) The tests must (by law) include laboratory tests in which <u>animals</u> are given the additive in <u>high doses</u>.

2) The tests are designed to give information about:
   - potential <u>side effects</u>
   - potential to cause <u>cancer</u>
   - potential to affect <u>reproductive</u> processes
   - potential to affect the development of an <u>embryo</u>

3) The <u>results</u> of safety tests are assessed by <u>independent experts</u>.

4) If an additive <u>passes</u> the safety tests, it is given an <u>E-number</u>, which means it is <u>approved</u> for use in the <u>European Union</u>.

Additives can't be used in food without having had vigorous testing.

So that explains how crisps can taste like spicy beef... Wow...

So that's what <u>E-numbers</u> are. I always wondered what the 'E' was all about. Still though, it worries me that they have to check about the chance of these additives causing cancer and weirdy side effects. If I thought there was any slight chance, I wouldn't be thinking of putting them in food, that's for sure.

Pros and Cons of Food Additives

You probably guessed from the last page that additives are great, but also, in some ways, worrisome. Here's more about the pros and cons of additives, and some case studies for you to talk about at parties.

Additives Have Advantages...

Additives are great because they can:

1) Improve the appearance, taste, flavour and texture of foods.
2) Add colour to food, or replace colours lost during processing.
3) Increase the shelf-life of food (i.e. make it last longer).
4) Improve or preserve the nutrient value of foods.

These lovely treats wouldn't look, smell or taste so nice if it weren't for additives.

...But They Have Disadvantages Too

Additives are not so great for a few reasons:

1) We're still not completely sure how safe some food additives are, especially in the long-term.
2) Some additives can cause allergic reactions in some people.
3) Some scientists think that some additives can cause hyperactivity in children, asthma, and even cancer — although more research is needed for us to be sure about this.

Here are Some Case Studies About the Health Risks of Additives

CASE STUDY 1: TARTRAZINE

1) Tartrazine is a yellow food colouring.
2) Scientific research has linked it to causing eczema, asthma, and hyperactivity in children.
3) Tartrazine has been banned in Norway, Austria and Finland because of this research.
4) However, it is very popular with the food industry in the UK and USA because it is cheaper than natural alternatives.

Here are some edible items with suspiciously artificial colours.

CASE STUDY 2: SACCHARIN

1) Saccharin is an artificial sweetener.
2) Research in the 1970s suggested that saccharin might cause cancer in laboratory animals.
3) A ban was proposed in America, but because saccharin was the only artificial sweetener available at the time, the ban was opposed by the diet food industry, the drinks industry and diabetics.
4) The research has since been criticised because of the huge doses used, and no further research has shown any clear health risks in humans at normal doses.

Where did the blue Smarties go? They were my favourite...

Blue Smarties, apparently, have been replaced with white ones — Smarties makers are now only using natural food colourings, which is great. It's pretty hard to avoid additives though, even if you try. No more frozen pizza, packet ham, fruit squash, cheese slices, fizzy drinks, sweets... You get the picture.

Interpreting Food Labels

There's a whole world of information on food labels about what's in the product, what you're meant to do with it, whether it's good for you etc...

Food Labels Have to Tell You Certain Information

Legally the label on processed food has to tell you the following information:

- The name of the product and what it is.
- How the product should be stored.
- A 'best before' or 'use by' date.
- What ingredients the product contains, in descending order of weight. Preservatives, colourants, emulsifiers and other additives are listed in the ingredients list.
- Whether a product contains genetically modified soya or maize ingredients.
- The name and address of the manufacturer.
- The weight or volume of the product.
- Instructions for preparation and cooking.

Food Labels Often Have Tables of Nutrients and Energy

Here's an example of a typical food label. You should be able to work stuff out using it. Like this:

1) How much of a nutrient is in a packet of food.

Question: A bodybuilder is monitoring how much protein he eats. How much protein is in the 40 g bar of 'Charlie's Chocolate' that he eats?

Answer: Mass of a nutrient in a packet (g)
= (mass of nutrient in 100 g ÷ 100) × mass of packet (g)
= (3 g ÷ 100) × 40 g = 1.2 g

2) The mass of a product that would give you a certain amount of energy.

Question: A dieter wants some chocolate, but only wants 200 kJ of energy from it. How much 'Charlie's Chocolate' could he eat?

Answer: Mass of food (g)
= (100 ÷ energy in 100 g) × chosen amount of energy
= (100 ÷ 1990 kJ) × 200 kJ = 10.1 g

The energy units here have to be the same — both in kJ, or both in kcal.

The amount of energy is given in kilojoules (kJ) or Calories (kcal).

Information is usually provided for 100 g of the product.

Charlie's Chocolate
Nutritional information:

	Typical values per 100 g
Energy	1990 kJ / 475 kcal
Protein	3 g
Carbohydrate	47 g
Fat	30 g
Fibre	7 g
Sodium	trace

Ingredients: sugar, cocoa mass, cocoa butter, butter fat, emulsifier (lecithin (soya)), flavouring.

Allergy advice: contains milk, soya, produced in a factory handling nuts.

The amount of each nutrient is given in grams (g).

There are Other Features on Some Labels

Here are some other bits and bobs that you might find on certain food labels:

1) Allergy advice — some foods are labelled with a warning that they contain ingredients that some people have allergic reactions to, e.g. nuts, milk, eggs or shellfish.

2) Suitable for vegetarians / vegans — vegetarian foods contain no meat or products resulting from the slaughter of animals, and vegan foods contain no meat or other animal products (e.g. milk and eggs).

3) Organic certification (e.g. Soil Association) — organic foods are produced without the use of artificial fertilisers or pesticides (for plants), or antibiotics or medicines (for animals).

4) Healthy eating logos — are used on foods with reduced levels of, e.g. calories, fat, salt or sugar.

5) Fairtrade logo — guarantees that food producers in the developing world received a fair price.

'This packet of walnuts may contain nuts'...

'This product may be hot when heated', 'Serving suggestion: defrost', 'Open packet, eat nuts', 'Do not turn upside down' (printed under the packet)... you've got to love food labels...

Qualitative Food Tests

It all gets a bit <u>technical</u> from here on in. The next three pages are all about the <u>tests</u> food scientists use to find out all the stuff on the previous pages about what's in what food.

1) The Iodine Test for STARCH — Turns It Blue/black

Here's what food scientists do to test for the presence of <u>starch</u> in food:

1) Add some drops of brown <u>iodine solution</u> to the food.

2) If the food contains <u>starch</u> the iodine will turn a <u>blue/black</u> colour.

Starch for sure

2) The Biuret Test for PROTEIN — Turns CuSO₄ Purple

To see whether there's any <u>protein</u> in food:

1) Add some <u>sodium hydroxide</u> (NaOH) solution to the food and <u>shake</u> with care.

2) Then add some <u>weak copper sulfate</u> ($CuSO_4$) solution.

3) If the <u>pale blue</u> colour turns <u>purple</u>, there's <u>protein</u> present.

Protein for sure

3) The Benedict's Test for SIMPLE SUGARS — a Brick-Red Precipitate

To test for <u>simple sugars</u>:

1) Add blue <u>Benedict's solution</u> to the food in a boiling tube. Then heat to <u>60 °C</u>.

2) If there's a <u>brick-red precipitate</u>, that means the food contains <u>simple sugars</u>, e.g. glucose, fructose, etc. (see p.48 for more details).

4) The Alcohol-Emulsion Test for FATS

Here's how to test for <u>fats</u>:

1) Mix the food with <u>pure ethanol</u> and then <u>filter</u> it.

2) Add the clear solution to <u>water</u>.

3) A <u>milky white emulsion</u> indicates that the food contains <u>fats</u>.

Milky white emulsion – fat for sure

5) The pH Test for Acidity

1) <u>Acidity</u> is measured on the <u>pH</u> scale, which goes from <u>0 to 14</u>.

2) Anything <u>less</u> than 7 is <u>acid</u>. Anything <u>more</u> than 7 is <u>alkaline</u>.

3) If something is <u>neutral</u> it has <u>pH 7</u> (e.g. pure water).

4) The <u>strongest acid</u> has <u>pH 0</u>. The <u>strongest alkali</u> has <u>pH 14</u>.

<u>Acidity</u> of food can be measured using an <u>indicator</u> — a <u>dye</u> that changes colour depending on whether it's put in an <u>acid</u> or an <u>alkali</u>. 'Universal indicator' is a <u>combination</u> of dyes which gives the colours shown below when put into food.

pH 0   1   2   3   4   5   6   7   8   9   10   11   12   13   14

ACIDS   NEUTRAL   ALKALIS

car battery acid, stomach acid | vinegar lemon juice | acid rain | normal rain | pure water | washing-up liquid | pancreatic juice | soap powder | caustic soda (drain cleaner)

The 'eating it' test for fat is another good one...

These tests weren't invented just to give you something to do in Science... They really are used in the <u>real world</u> to find out the <u>properties</u> of different foods. How else do you think we know that beans contain protein and potatoes contain starch? They don't just grow with food labels on you know...

Quantitative Food Tests

Quantitative food tests tell us the __amount__ of different things in food. This information is needed to label foods correctly, and to make sure that foods are __safe__ and satisfy __legal requirements__. Important stuff.

Moisture Content __Can be Tested by__ __Evaporation__

The __moisture content__ of food affects __bacterial growth__ and __shelf-life__. The __more__ moisture, the more likely it is that bacteria will grow and the food will __spoil__. There are __legal standards__ for moisture content of many foods, especially things that should be __dry__ like grains, cereals, flour, sugar, coffee, tea and spices.

Here's how food scientists __test__ the moisture content of a food sample:

1) Measure and record the __initial mass__ of the food sample.

2) __Heat__ the food sample in a warm __oven__ until the mass __stops decreasing__. (Take the sample out of the oven a few times to check its mass.)

All the water in the food will evaporate.

3) When the mass has stopped decreasing, take the food out of the oven and record the __final mass__.

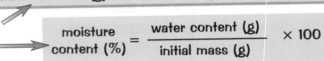

water content (g) = initial mass (g) − final mass (g)

4) Work out the __water content__ of the sample like this:

5) Then calculate the __moisture content__ like this:

$$\text{moisture content (\%)} = \frac{\text{water content (g)}}{\text{initial mass (g)}} \times 100$$

Suspended Matter __Can be Tested for by__ __Filtration__

__Suspended matter__ in liquids makes them look __cloudy__ and can show that __bacteria__ are present. There is a __legal limit__ for the amount of suspended matter in tap water. In fruit juice, most suspended matter is fruit, so juices are tested for suspended matter to ensure they're __good quality__.

Here's how drinks can be __tested__ for suspended matter:

1) Measure out 100 ml of the sample of drink.

2) __Filter__ the sample using a __conical flask__, __filter funnel__ and __filter paper__.

3) Carefully __dry__ the filter paper.

4) Measure and record the __mass__ of the __residue__ (the stuff left on the filter paper).

5) Work out the __suspended matter content__ of the sample in __grams per litre__ (__g/l__) by multiplying the mass of the residue by 10. For example, 0.82 g of residue in 100 ml gives a suspended matter content of 8.2 g/l.

filter paper — *residue*

filtered orange juice

Acidity __Can be Tested for by__ __Titration__

The __acidity__ of many foods and drinks affects their __quality__ and __shelf-life__ and needs to be __tested__. For example, the acidity of __jam__ is tested during its production to make sure it will __set__ properly, and the acidity of __vinegar__ is tested to ensure that it's __safe__ to eat and will __keep well__. Here's how to __test__ vinegar's acidity:

1) __Pipette__ 25 cm³ of vinegar into a beaker, and make it up to 250 cm³ with __distilled water__.

2) Pipette 25 cm³ of this diluted vinegar solution into a __conical flask__ and add 3-4 drops of __pH indicator__.

3) Fill a __burette__ with 0.1 M __sodium hydroxide solution__.

4) Place the flask of diluted vinegar under the burette, and add the sodium hydroxide solution __drop by drop__, until the __colour__ changes permanently. Record the __volume__ of sodium hydroxide used (in cm³).

5) __Repeat__ the titration twice and __average__ your results.

6) This value multiplied by 0.24 gives the __concentration__ of __ethanoic acid__ in the undiluted vinegar (in g per 100 cm³). Exactly what __number__ you multiply by depends on the __acid__ you're testing.

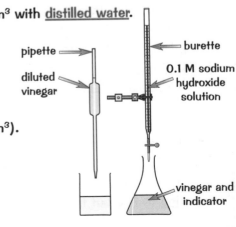

pipette — *burette*

diluted vinegar — *0.1 M sodium hydroxide solution*

vinegar and indicator

Quantitative Food Tests

Don't go to sleep just yet... here are two more riveting food tests for you to learn about. Great.

Vitamin C Content Can be Tested for by Titration

Many foods, food supplements and drinks are tested for <u>Vitamin C</u> content — so that the <u>nutritional information</u> can be labelled correctly.

Here's how to <u>test</u> for vitamin C:

1) Prepare an '<u>extract</u>' of the food sample to be tested — <u>grind</u> the sample, add <u>water</u>, and then <u>filter</u> it to leave a liquid food extract.

2) <u>Pipette</u> 1 cm³ of the food extract into a test tube.

3) Clamp the test tube in place under a <u>burette</u> containing 0.1% <u>DCPIP</u> solution (DCPIP is a <u>blue dye</u> that turns colourless in vitamin C).

4) Add the DCPIP <u>drop by drop</u>. When the first drop is added, the DCPIP's <u>blue</u> colour should quickly <u>disappear</u>. Keep adding DCPIP until the blue colour <u>doesn't</u> disappear any more, and note the <u>volume</u> of DCPIP used.

5) <u>Repeat</u> the titration twice and <u>average</u> your results.

6) The volume of DCPIP used can be used to <u>work out</u> the volume of <u>vitamin C</u> present in the extract.

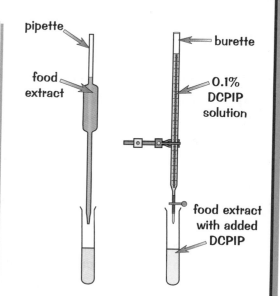

Iron Content Can be Tested for Using Standard Solutions

Many foods, food supplements and drinks are tested for <u>iron content</u> so that the <u>nutritional information</u> can be labelled correctly. Here's how they do it:

1) Dilute 1% <u>iron(III) chloride</u> solution to obtain solutions with concentrations of 0.1%, 0.01%, 0.001%, 0.0001%, 0.00001% and 0.000001%.

2) Pour 5 ml of each solution into a separate test tube, and add 5 ml of 0.1 M <u>KSCN</u> solution to each tube. (KSCN is a chemical that turns the solution a <u>deeper red</u> colour the <u>more iron</u> that's in it.) These are your <u>standard solutions</u>.

Meanwhile...

3) Measure and record the <u>mass</u> of your <u>food sample</u>.

4) <u>Heat</u> the food sample strongly in an evaporating dish or crucible until all that's left is a <u>grey ash</u>.

5) After <u>cooling</u>, add 5 ml of <u>distilled water</u> to the ash, stir well, and <u>filter</u> into a test tube.

6) Add 5 ml of 0.1 M <u>KSCN</u> solution to the test tube, and <u>compare</u> the <u>colour</u> to the standard solutions.

7) The standard solution that's the most <u>similar</u> in <u>colour</u> to the food sample will have roughly the same <u>concentration</u> of iron as the food sample.

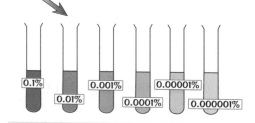

Here's what you get after heating and cooling your food sample, mixing it with water, filtering it and then adding KSCN solution. <u>Compare</u> it to the standard solutions above — I think this one has an iron content of about <u>0.01%</u>.

So that stuff they make railways from is in my food...?

That's the end of that then. It wasn't so bad after all was it? Those food tests take a bit of learning, but if you've had a go at them in your Science classes, you'll find them easier to remember. Anyway, it's all good meaningful stuff and you should learn it for your own health as well as those exams. So there.

Revision Summary for Section 2.1

Well done — you've made it to the end of the section. I knew you could do it. But wait — it's not time to put the book down and have a bag of crisps. That would just prove how little you'd learned from this section. At the very least you should be going off to have a nice fruit salad. But you shouldn't be going off anywhere to eat anything until you've discovered how much of this stuff you've taken in.

And what better way to do it than with a lovely page of questions. Mmmm...

1) Name the five main types of nutrient and say what their functions are in the body.
2) List the nutrients you'd be getting if you ate meat, eggs, pasta and fruit for dinner.
3) What exactly is fibre and why do we need it in our diets?
4) List the types of food in which you'd find most fibre.
5) Name five vitamins and say why our bodies need them.
6) Which vitamin is produced by the body when exposed to sunlight?
7) In which type of food would you find both vitamins A and K?
8) Name four minerals and state their functions in the body.
9) What is likely to happen to you if you don't get enough vitamin A?
10) Bleeding gums and poor healing of wounds is a symptom of deficiency of which vitamin?
11) What could happen to you if you have too much saturated fat or salt in your diet?
12) Name a disorder associated with too much sugar in the diet.
13) Which two units are used to measure the amount of energy in food?
14) Give a good reason why we shouldn't live on a diet of chocolate cake and lard.
15) What is 'fast food'? Why should we try to reduce the amount of fast food we eat?
16) Why have some fast food companies got into trouble for aiming their marketing at children?
17) Name the six main types of food additive, and give an example of each.
18) Which additives are likely to be found in:
    a) fizzy drinks,   b) sweets and   c) low-calorie ice cream?
19) What potential side effects might additives be tested for?
20) Give three advantages and three disadvantages of using food additives.
21) What is tartrazine, what health problems has it been linked to and why is it still being used?
22) What is saccharin, what disease was it linked to and who opposed its ban?
23) List eight pieces of information that must be given on the labels of processed food.
24) Using the information on a food label, describe how you would work out the number of grams of fat in 60 g of the food.
25) List five other things that you might find on a food label.
26) How could you test for the presence of starch, protein or simple sugars in a sample of food?
27) If you were using the alcohol-emulsion test, what would indicate the presence of fats?
28) If you're testing pH using universal indicator, what would it mean if the dye turned orange?
29) Briefly describe how you could test the moisture content of a sample of food.
30) What can suspended matter in water signify?
31) Briefly describe the process of titration. Name two things titration can be used to test foods for.
32) When testing for vitamin C, what solution is added drop by drop to the extract?
33) When testing for iron content, what are 'standard solutions' and how are they used?

Microorganisms in Food Production

Microorganisms such as bacteria, yeast and other fungi are really useful in food production. It sounds a bit yucky but there are plenty of foods that wouldn't taste nearly as nice without the help of bacteria and mould. Microbiologists study these microorganisms to work out exactly how we can use them to make useful products. They also investigate what factors affect the growth of microorganisms — this allows us to control conditions to get the best results in food production.

Bacteria are Used to Make Cheese...

Believe it or not, cheese doesn't come straight out of cows ready-packaged (that'd just be weird) — it's made by adding bacteria to milk like so:

1) A culture of bacteria is added to warm milk.
2) The bacteria produce solid curds in the milk.
3) The curds are separated from the liquid whey.
4) More bacteria are added to the curds, and the whole lot is left to ripen for a while.
5) Moulds are added to give blue cheese (e.g. Stilton) its colour and taste.

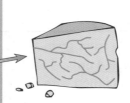

This woman is cutting up the solid curds so that she can drain off the liquid whey.

...and Yoghurt

Here's another surprise for you — tasty yoghurt is also made from milk using bacteria. Here's how:

1) All the equipment is sterilised to kill off any unwanted microorganisms.

2) Then the milk is pasteurised (heated up to 72 °C for 15 seconds) — again to kill off any unwanted microorganisms. Then the milk's cooled.

3) A starter culture of bacteria is added to the milk. The mixture is then incubated (heated to about 40 °C) in a vessel called a fermenter. The bacteria convert the lactose sugar in the milk into lactic acid — this is called fermentation. The lactic acid causes the milk to clot and solidify, turning it into yoghurt.

4) A sample is taken to make sure it's at the right consistency. Then flavours (e.g. fruit) and colours are sometimes added and the yoghurt is packaged.

Anyone for a nice pot of acidy milk, mmmmm...

The world's fastest yoghurt — pasteurised before you see it...

Not all microorganisms are bad for you — some of them can help make some really tasty food, like cheese. You still need to stop the bad microorganisms infecting your food though. That's why the milk is pasteurised — heating it kills off microbes, making it perfectly safe and ready for your enjoyment.

Microorganisms in Food Production

Yeast is a pretty handy fungus — it helps us make bread, beer and wine...

Yeast is Used to Make Bread

Holes in the bread, which make it nice and light, are made by carbon dioxide bubbles in the dough.

1) Yeast is used in dough to produce nice, light bread.
2) The yeast converts sugars to carbon dioxide and some ethanol. It's the carbon dioxide that makes the bread rise.
3) As the carbon dioxide expands, it gets trapped in the dough, making it lighter.

Brewing Beer and Wine Also Needs Yeast

There are four main steps to brewing beer and wine...

1 Firstly you need to get the sugar out of the barley or grapes:

BEER Beer is made from grain — usually barley. The grains are mashed up and soaked in water to produce a sugary solution with lots of bits in it. This is then sieved to remove the bits.

WINE Wine is made from grapes. These are mashed and water is added.

2
- Yeast is added and the mixture is incubated (heated up). The yeast converts the sugar into alcohol — this is another type of fermentation.
- The mixture is kept in fermenting vessels designed to stop unwanted microorganisms and air getting in.
- The rising concentration of alcohol in the mixture eventually starts to kill the yeast. As the yeast dies, fermentation slows down.

3 The beer and wine produced is drawn off through a tap. Sometimes chemicals called clarifying agents are added to remove particles and make it clearer.

4 The drink is packaged ready for sale:

BEER The beer is usually then pasteurised — heated to kill any yeast left in the beer and completely stop fermentation. Finally the beer is casked ready for sale.

WINE Wine isn't pasteurised — any yeast left in the wine carries on slowly fermenting the sugar. This improves the taste of the wine. The wine is then bottled ready for sale.

Always invite mushrooms to parties — they're fun guys...

There are lots of fancy things you can do with beer which varies its taste — hops are usually added to the mixture to give the beer its bitter flavour. The type of yeast that's used also affects the end product — lager uses a different yeast to ale, which makes it all bubbly (don't ask me how).

Food Poisoning

If you've ever had <u>food poisoning</u> you don't need me to tell you how gruesome it is. All that regurgitated food just isn't pretty. So read this page and make sure you know <u>what it is</u> and <u>what causes it</u>.

Food Poisoning is Caused by Microorganisms in Food

1) Food poisoning is usually caused by the presence of <u>microorganisms</u> (usually <u>bacteria</u>) in food.

2) These can make you ill if you ingest them, by <u>directly harming</u> your cells, or by <u>producing toxins</u> that poison you.

3) There are many different ones, but common problem bacteria are:
   - <u>Campylobacter</u> — found in dairy and poultry.
   - <u>Salmonella</u> — found in poultry and eggs.
   - Types of <u>E.coli</u> — usually found in raw meat.

4) Food poisoning can occur when food isn't <u>stored</u> or <u>cooked properly</u>, through <u>poor kitchen</u> or <u>personal hygiene</u> or <u>contamination</u> from other sources.

The Symptoms Depend on the Type of Microorganism

The <u>type of microorganism</u> that you're infected with can affect:

1) The <u>symptoms</u> you get — these can include:
   - <u>stomach pains</u>
   - <u>vomiting</u>
   - <u>diarrhoea</u>

2) <u>How ill</u> you get, <u>how long</u> it lasts and <u>the time</u> it takes to start from when you've eaten the contaminated food.

Food Can Become Contaminated

In <u>mass food production</u> it's hard to prevent food from becoming contaminated.

1) Most of the contaminants are harmless, but some can cause <u>problems</u> e.g. plastic, glass, metal, banned additives, microorganisms and insects.

2) If food is <u>contaminated</u> people can get <u>ill</u> or <u>injured</u>. The manufacturer will have to <u>recall the product</u>, which can be <u>expensive</u> and <u>damaging</u> to the company's reputation.

3) Manufacturers must also tell you on the <u>label</u> exactly what's in (or <u>could</u> be in) the food, especially if any ingredients (e.g. peanuts or wheat) could trigger <u>allergies</u>.

> **AN EXPENSIVE FOOD RECALL** Routine tests by Public Health Inspectors found that a sample of Worcester sauce contained <u>Sudan 1</u>, a banned food colour. This was traced back to a batch of <u>chilli powder</u> imported from India. A massive search took place to find all other products that had used this Worcester sauce or chilli powder in their manufacture. The food companies involved had to remove the contaminated products from shops and inform their customers and the Food Standards Agency of the mistake. The incident cost them a fortune and may have damaged their reputation.

It was the salmon mousse I tell you...

Some <u>bacterial toxins</u> are amongst the most <u>potent poisons</u> known to man. <u>Botox</u> is actually a bacterial toxin. If you eat the bacteria that produce the toxin you can get <u>botulism</u> (a type of food poisoning). If you inject the toxin into your face you'll end up looking like Elizabeth Taylor. Hmmm.

Food Hygiene

Right, time to get your rubber gloves out — this page is all about being hygienic. You might not think that cleanliness is next to godliness but you will once you've spent an uncomfortable night inspecting the U-bend.

There are Five Main Ways to Keep Kitchens Hygienic

① KEEP YOURSELF CLEAN

To make sure you're free of bacteria when handling food, you should:

1) Wash your hands — especially before preparing food, after handling raw meat, going to the toilet and after touching rubbish bins. It's not just fussiness — it's because microbes can stay alive on your hands for ages and can spread to everything that you touch.

2) Wear a hairnet or hat — just one hair can carry 100 000 microbes.

3) Cover all cuts and wounds to stop bacteria getting into the cuts and to stop any bacteria in the cuts getting into food.

4) Remove jewellery, e.g. rings, watches and bracelets — there could be millions of microbes underneath them. Urggh.

② USE DISINFECTANTS AND DETERGENTS

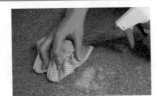

1) Disinfectants (e.g. antibacterial cleaners and bleach) are chemicals that kill microbes. They're used for cleaning work surfaces and floors.

2) Detergents (e.g. washing-up liquid and washing powders) dissolve grease, oil and dirt. They deprive the microorganisms of the food they need to live.

Things can be sterilised using heat, disinfectant or radiation.

③ USE HEAT TO STERILISE EQUIPMENT

Heat can be used to kill all the microbes on equipment, steam is often used to do this. In the home we use steam to sterilise things like babies' bottles.

④ DISPOSE OF WASTE PROPERLY

Kitchen bins are like a giant breeding ground for microorganisms and they attract pests like rats and cockroaches to top it all off. So, they should be emptied regularly and should have a lid and a liner.

⑤ CONTROL PESTS

Animals aren't allowed anywhere near food preparation areas — they can carry harmful microorganisms. Here's how to deal with the common pests...

1) Mice and rats can be trapped or poisoned.

2) Cockroaches and other crawling insects can be trapped with glueboard traps (a sticky plastic sheet left out in areas where they are present).

3) Flying insects can be zapped with ultraviolet light traps — they're attracted to the light and fly into them, hitting an electric grid that kills them. You can stop them getting into the building in the first place by putting fine mesh screens over windows and chain curtains across doors.

Control pests — tie up your sister...

In industry these five things are really important to make sure food is safe to eat. Hygiene standards are checked by Public Health Inspectors. It's still important to do these things at home though (I'm not suggesting you don't... you look like a very clean person — I bet your mother's really proud).

Food Preservation

Food manufacturers <u>don't</u> want to give their customers <u>food poisoning</u> — that'd be pretty bad for business. The manufacturers <u>change the conditions</u> in the food to <u>reduce bacterial growth</u> (bacteria are a bit fussy about what conditions they'll grow in and if they're not right they won't grow — stubborn things).

Bacteria Like Warm, Moist Conditions

Bacteria can pretty much exist anywhere. The ones that make food go off and make you ill prefer:

- <u>Warmth</u> — this helps the <u>reactions</u> in bacteria to go faster, but only up to a certain point. They generally need temperatures from <u>5 to 62 °C</u>.
- <u>Moisture</u> — bacteria need <u>water</u> to survive.
- <u>Food source</u> — like all other organisms, they need food to grow. Greedy little blighters.
- Many bacteria also like a <u>neutral pH</u> (6.5 to 7.5).

The Growth of Bacteria Can be Slowed Down or Stopped

Depriving bacteria of the conditions they need to grow, will either <u>slow</u> or <u>stop</u> their growth.

REFRIGERATION

Keeping food <u>below 5 °C</u> slows down the growth of any bacteria present in the food. It's just <u>too cold</u> for their reactions to work fast enough. Brrr.

FREEZING

In the freezing process, the <u>moisture</u> which bacteria need to thrive is <u>frozen</u>, so the bacteria can't grow and multiply. Freezers operate at around <u>−18 °C</u> (domestic) or <u>−32 °C</u> (industrial) — that's <u>far too cold</u> for most bacteria.

HEATING

Cooking food <u>kills bacteria</u> (as long as it's cooked right through). <u>Pasteurisation</u> is used in industry — food is heated to a <u>high temperature</u> to destroy any bacteria present. It's then very <u>carefully packaged</u> to make sure no bacteria (or anything else for that matter) can contaminate it.

DRYING

<u>Drying</u> removes all the <u>moisture</u> so bacteria can't grow. <u>Accelerated freeze drying</u> is another form of drying — food is <u>frozen</u> and <u>dried</u> in a <u>vacuum</u>. The quick freezing process means that the food keeps its <u>colour</u> and <u>flavour</u> (it's used for things like coffee and packet soups).

SALTING

Adding salt means the bacteria <u>can't take in water properly</u>, stopping them from growing. Salt can be added to <u>meat</u> to make it last longer.

PICKLING

This involves storing food in <u>vinegar</u>, which is acidic. Bacteria don't like the <u>acidic</u> conditions so won't grow very fast.

Mmmm... the nicest pickled onions you've ever seen.

If you don't learn this you'll be in a pickle...

Food manufacturers have to do this stuff to <u>stop</u> their food <u>going off</u> before it's on the shop shelves. It applies to home life too — some food needs refrigerating and you need to cook food properly.

Detecting Bacteria

Food products are regularly tested for bacteria. This might be by a <u>health inspector</u> or by the <u>company itself</u> (to make sure their food is of a consistent quality). Food preparation areas and processing equipment are tested too.

Food Products are Tested for Bacteria...

1) <u>Public Health Inspectors</u> routinely test food from hotels, fast food outlets and supermarkets for bacteria to make sure <u>food is safe</u> and to <u>prevent outbreaks</u>.

2) If an outbreak of <u>food poisoning</u> does occur, food will be tested to find the <u>source</u>.

3) A sample of food is tested for:

- The <u>amount</u> of bacteria in the food.

- The presence of any <u>harmful bacteria</u>, e.g. Salmonella.

4) Certain <u>levels</u> of bacteria are <u>acceptable</u> — it's impossible to get rid of them all.

See the next page for how they do this.

...so are Equipment and Surfaces

1) In food production everything that food might come into <u>contact</u> with must be <u>clean</u>. This includes <u>equipment</u> as well as <u>work surfaces</u>.

2) <u>Regular checks</u> are carried out to ensure nothing harmful is lurking around which might <u>contaminate</u> the food during <u>preparation</u>.

3) Samples are taken from surfaces by <u>swabbing areas</u> using swabs (a bit like cotton buds). The swabs will then be taken to a lab to be <u>tested</u>.

You Use Aseptic Techniques to Prevent Contamination

<u>Aseptic techniques</u> are standard procedures used by microbiologists to <u>prevent contamination</u> — they should be used when <u>collecting</u> and <u>analysing</u> samples. This is done so you know that any bacteria you find <u>came from the food sample</u> (or surface or equipment) you were testing, not from a dirty lab bench or a grubby Petri dish etc. This is what you should do:

1) <u>Sterilise</u> all equipment <u>before</u> and <u>after</u> use.

2) Keep samples containing microorganisms in <u>sample bottles with lids</u>.

3) When opening a sample bottle to use it, <u>close it</u> again <u>as soon as possible</u>.

4) <u>Pass</u> the tops of sample bottles through a <u>Bunsen flame</u> whenever lids are <u>removed</u>.

5) <u>Don't</u> put lids down on <u>benches</u> — hold them with your <u>little finger</u> or your <u>other hand</u>.

6) Don't open Petri dishes <u>until</u> you are <u>ready to use them</u>.

7) Don't put any <u>equipment</u> that comes into <u>contact with microorganisms</u> down on <u>benches</u>.

8) <u>Seal</u> agar plates with sticky tape and <u>label</u> them with your <u>name</u>, the <u>date</u> and <u>what</u> you've put on the plate.

9) <u>Don't open</u> agar plates once they have been <u>sealed</u>.

10) <u>Dispose</u> of cultures <u>safely</u> — usually done by pressure sterilising in an <u>autoclave</u>.

An agar plate is a Petri dish containing a jelly-like substance called agar.

Don't forget you need to wear <u>gloves</u> and <u>protective clothing</u> when dealing with microbes. This stops microbes from you contaminating your sample and protects you from infection at the same time.

There are more bacteria on the kitchen sink than in the toilet...

All this aseptic technique stuff might seem like a lot to remember but it's all the same basic principle — be as <u>clean</u> and <u>safe</u> as possible, and don't leave things lying around with lids off. There are lots of bacteria in the air which could easily fall into your agar, so make sure you catch as few as possible.

Detecting Bacteria

The samples taken from surfaces, equipment or food are then analysed in a microbiology lab (by a microbiologist) to find out how many bacteria are present and what type they are. Sounds thrilling...

The Serial Dilution Method is Used to Count Bacteria

Serial dilutions are used to calculate how many bacteria are present in a sample — the original sample will contain too many bacteria for someone to count, so you dilute the sample again and again until there are few enough for you to count them. Here's how it's done:

1) You start by diluting your sample by mixing a little bit of it with water. Then it's diluted again and again etc.

2) A small amount of the final dilution is spread over an agar plate and incubated. Each bacterial cell found in this final dilution will reproduce to produce a visible colony (a clump of bacteria).

3) The number of colonies is counted. If you know how many times the sample was diluted (and by how much, e.g. by a factor of 10), you can work out the approximate total number of cells in the original sample.

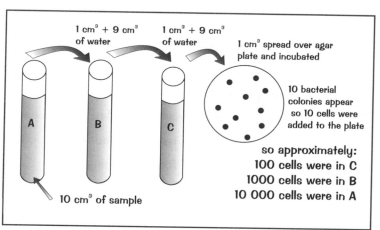

1 cm³ + 9 cm³ of water   1 cm³ + 9 cm³ of water   1 cm³ spread over agar plate and incubated

A   B   C

10 cm³ of sample

10 bacterial colonies appear so 10 cells were added to the plate

so approximately:
100 cells were in C
1000 cells were in B
10 000 cells were in A

You Make Streak Plates to Isolate Bacteria for Identification

If you take a swab from a surface or test food for bacteria it makes sense for you to find out what type of bacteria are present — so you know if there are any harmful ones present. Here's how you do it:

1 Flame a wire inoculating loop to get rid of any bacteria.

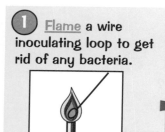

2 Dip the wire into the sample.

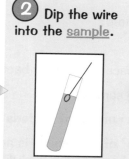

3 Using the wire loop spread the broth over an agar plate, so the bacteria are spread out as much as possible. Then incubate the agar plates (keep them warm) so the bacterial colonies grow.

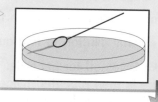
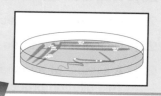

You need to use aseptic techniques when doing streak plates (and serial dilutions) because you need to know that the microorganisms that you've grown come from the sample (not from a dirty lab bench or from the air).

4 A colony can be identified by removing it from the plate and staining the bacteria. Then you use a microscope to identify the type of bacteria.

Cereal dilutions — adding too much milk...

What the health inspectors do next depends on the results of these tests. If they find that a food sample contains way too many bacteria or a particularly harmful bacterium they might recall the product. If they find the same thing on surfaces or equipment they might shut down the restaurant, processing plant etc.

Revision Summary for Section 2.2

¡Hola! y bienvenido a Quién quiere ser millonario. Responda correctamente a quince preguntas y usted podría volver a casa con un millón de euros. Juguemos. Oh hang on, wait a minute, I forgot I'm not presenting the Spanish Who Wants To Be A Millionaire any more, sorry. Oh how the mighty have fallen. You don't have any lifelines, and the chances of winning a million by answering these questions are mighty slim, but why not give them a go anyway.

1) Describe the process of making cheese.

2) Describe the process of making yoghurt.

3) In yoghurt making, why is the milk pasteurised?

4) Why is yeast used when making bread?

5) Describe the four main steps in brewing beer and wine.

6) What does yeast convert the sugar into?

7) What type of chemical might be added to the beer or wine to make it clearer?

8) Name three types of bacteria that cause food poisoning.

9) Give three symptoms of food poisoning.

10) Give six things that food can become contaminated with.

11) What four things should you do to make sure you are clean before you start to prepare food?

12) What are disinfectants used for?

13) Why should you dispose of kitchen waste properly?

14) How could you control flies in a kitchen?

15) Give four conditions most harmful bacteria prefer.

16) Describe five ways you can stop or slow down bacterial growth.

17) Give two examples of foods that are preserved.

18) Why do health inspectors regularly test food products?

19) What two things are food samples tested for?

20) Why are food processing equipment and work surfaces tested for bacteria?

21) Why are aseptic techniques important when taking samples?

22) Describe how to do a serial dilution to find out the number of bacteria in a sample.

23)* A sample of milkshake is tested for bacteria. 1 cm^3 of the sample is added to 9 cm^3 of water. This dilution procedure is then repeated twice and 1 cm^3 of the final dilution is spread on an agar plate and incubated. Six colonies grow. Approximately how many bacteria were present in the original 1 cm^3 sample?

24) Describe how to make a streak plate to isolate a bacterium for identification.

* Answer on page 100.

Essential Nutrients

Plants need various <u>minerals</u> from the soil, as well as the <u>carbohydrates</u> they make by <u>photosynthesis</u>, so that they can grow big and healthy, and then, after all that hard work, we eat them.

Plants Need Minerals for Healthy Growth

1) Plants need certain <u>elements</u> so they can produce important compounds.

2) They get these elements from <u>minerals</u> in the <u>soil</u>.

3) Sometimes the plant <u>can't</u> get all of the mineral ions it needs to be healthy. It depends what's there in the <u>soil</u> — if the supply in the soil gets <u>low</u>, the plant can't just wander off and find some more. It has to put up with it, and eventually it will start to show <u>deficiency symptoms</u>.

4) If the plant is left short of the minerals it needs for a long time, it might <u>die</u>.

1) Nitrates

Nitrates are needed to make proteins, which are needed for <u>cell growth</u>.

If a plant can't get enough nitrates it will be <u>stunted</u> and will have <u>yellow older leaves</u>.

2) Phosphates

Phosphates are needed for <u>respiration</u> and <u>growth</u>.

Plants without enough phosphate have <u>poor root growth</u> and <u>purple older leaves</u>.

3) Potassium

Potassium is needed to help <u>photosynthesis</u> and <u>respiration</u>.

If there's not enough potassium in the soil, plants have <u>poor flower and fruit growth</u> and <u>discoloured leaves</u>.

4) Magnesium (in smaller amounts)

Magnesium is essential for making <u>chlorophyll</u> (needed for <u>photosynthesis</u>).

Plants without enough magnesium have <u>yellow leaves</u>.

The Minerals in Soil Need to be Replaced

1) Plants take <u>essential nutrients</u> from the soil in order to <u>grow</u> and <u>reproduce</u>.

2) When plants <u>die</u> they are broken down by <u>microbes</u> — so the nutrients are <u>returned</u> to the soil, ready for more plants to use.

3) But if the <u>plants</u> are <u>taken away</u> by the farmer (for us and other animals to eat) then the <u>nutrients</u> are also <u>taken away</u>.

4) This means the minerals in the soil aren't <u>naturally replaced</u>, and the farmer has to use other methods to replace them, e.g. manure, compost or chemical fertilisers.

Nitrates and phosphates and potassium, oh my...

When a farmer or a gardener buys <u>fertiliser</u>, that's pretty much what they're buying — a nice big bag of mineral salts to provide all the extra elements plants need to grow. The one they often need most of is <u>nitrate</u>, which is why manure works quite well — it's full of nitrogenous waste excreted by animals.

Intensive Farming

Farmers use what they know about plants to farm more <u>efficiently</u>, and with the world's <u>increasing</u> <u>population</u>, intensive farming could be just what we need — <u>bigger</u> and <u>better</u> yields.

Intensive Farming — Getting the Most out of Plants and Animals

1) <u>Intensive farming</u> is where farmers try to get <u>as much as possible</u> from their plants and animals.

2) The aim is to produce the maximum amount of <u>food</u> from the <u>smallest possible</u> amount of <u>land</u>, to give a <u>huge variety</u> of <u>quality</u> foods, <u>all year round</u>, at <u>cheap prices</u>. They do this in <u>three</u> main ways...

Intensive Farming Uses Artificial Fertilisers...

1) Plants need <u>certain elements</u>, e.g. <u>nitrogen</u>, <u>potassium</u> and <u>phosphorus</u>, so they can make important compounds like proteins.

2) If plants don't get enough of these elements, their <u>growth</u> and <u>life processes</u> are affected (see previous page).

3) Sometimes these elements are <u>missing</u> from the soil because they've been used up by a <u>previous crop</u>.

4) Farmers use artificial fertilisers to <u>replace</u> these missing elements or provide <u>more</u> of them. This helps to increase the <u>crop yield</u>.

...As Well As Pesticides, Fungicides and Herbicides

1) <u>PESTICIDES</u> are chemicals that kill <u>farm pests</u>, e.g. insects, rats and mice. Pesticides that kill insects are called <u>insecticides</u>. Killing pests that would otherwise eat the crop means there's more left for us.

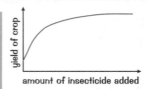

2) <u>FUNGICIDES</u> kill fungi, e.g. moulds that can damage crops.

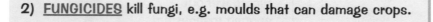

3) <u>HERBICIDES</u> kill <u>weeds</u>. If you <u>remove</u> plants that compete for the same <u>resources</u> (e.g. nutrients from the soil), it means the crop gets more of them and so grows better.

Animals can be Kept in Controlled Environments

1) In countries like the UK, animals such as <u>pigs</u> and <u>chickens</u> are often <u>intensively farmed</u> (battery farming).

2) They're kept <u>close together indoors</u> in small pens, so that they're <u>warm</u> and <u>can't move about</u>.

3) This saves them <u>wasting energy</u> on movement, and stops them giving out as much energy as <u>heat</u>.

4) This means the animals will <u>grow faster</u> on <u>less food</u>.

5) This makes things <u>cheaper</u> for the farmer, and for us when the animals finally turn up on supermarket shelves.

Intensive farming might just crop up in the exam...

The important stuff is knowing <u>how</u> intensive farming <u>increases</u> the amount of food — <u>fertilisers</u> provide essential <u>minerals</u> for growth, <u>herbicides</u> remove <u>competition</u>, <u>fungicides</u> prevent <u>disease</u>, and putting animals in <u>controlled environments</u> means they <u>waste less energy</u> (all the more for us).

Organic Farming

Intensive farming methods are still used a lot. But people are also using organic methods more and more.

Organic Farming Doesn't Use Artificial Chemicals

An alternative to modern intensive farming is organic farming. Organic methods are more traditional. Where intensive farming uses chemical fertilisers, herbicides and pesticides, organic farming has more natural alternatives.

THE LAND IS KEPT FERTILE BY:

1) Using organic fertilisers (i.e. animal manure and compost). This recycles the nutrients left in plant and animal waste. It doesn't always work as well as artificial fertilisers, but it's better for the environment.

2) Crop rotation — growing a cycle of different crops in a field each year. This stops the pests and diseases of one crop building up, and means nutrients are less likely to run out (as each crop has different needs).

PESTS AND WEEDS ARE CONTROLLED BY:

1) Weeding — physically removing the weeds, rather than just spraying them with a herbicide. Obviously it takes a lot longer, but there are no nasty chemicals involved.

2) Varying crop growing times — farmers can avoid the major pests for a certain crop by planting it later or earlier in the season. This means they won't need pesticides.

3) Using natural pesticides — some pesticides are completely natural, and so long as they're used responsibly they don't mess up the ecosystem.

4) Biological control — Biological control means using a predator, a parasite or a disease to kill the pest, instead of chemicals. For example:

a) Aphids are pests which eat roses and vegetables. Ladybirds are aphid predators, so people release them into their fields and gardens to keep aphid numbers down.

b) Certain types of wasps and flies produce larvae which develop on (or in, yuck) a host insect. This eventually kills the host. Lots of insect pests have parasites like this.

c) Myxomatosis is a disease that kills rabbits. In Australia the rabbit population grew out of control and ruined crops so the myxoma virus was released as a biological control.

Organic Farms Keep Animals in More Natural Conditions

1) For an animal farm to be classified as "organic", it has to follow guidelines on the ethical treatment of animals.

2) This means no battery farming — animals have to be free to roam outdoors for a certain number of hours every day.

3) Animals also have to be fed on organically-grown feed that doesn't contain any artificial chemicals.

Don't get bugged by biological pest control...

The Soil Association is an organisation that certifies farms and products as organic. They have very strict rules about what products can carry their logo — much stricter than the Government's minimum standards. About 70% of all organic food sold in the UK is Soil Association approved.

Comparing Farming Methods

It's all very well knowing how the different farming methods <u>work</u>, but are they actually any <u>good</u>? Both intensive and organic methods have <u>advantages</u> and <u>disadvantages</u>, which I'm afraid you just have to <u>learn</u>.

Intensive Farming is Efficient but can Damage the Environment

The main advantage of intensive farming methods is that they produce large amounts of food in a very <u>small space</u>, and it's <u>cheaper</u> for us. But they can cause a few <u>problems</u> — the main effects are:

1) <u>Removal of hedges</u> to make huge great fields for <u>maximum efficiency</u>.
   This <u>destroys</u> the <u>natural habitat</u> of <u>wild creatures</u> and can lead to serious <u>soil erosion</u>.
2) Lots of people think that intensive farming of <u>animals</u> such as <u>battery hens</u> is <u>cruel</u>.
3) The <u>crowded</u> conditions in factory farms create a favourable environment for the <u>spread of disease</u>.
4) If <u>too much</u> fertiliser is applied, it can find its way into rivers and lakes, causing the <u>death</u> of many <u>fish</u>.
5) <u>Pesticides</u> can build up to <u>toxic</u> levels in animals like otters and birds of prey.

Organic Farming Has Advantages and Disadvantages Too

ADVANTAGES

1) Organic farming uses fewer <u>chemicals</u>, so there's less risk of toxic chemicals remaining on food.
2) It's better for the <u>environment</u>. There's less chance of <u>polluting rivers</u> with fertiliser.
   Organic farmers also avoid using <u>pesticides</u>, so don't disrupt food chains and harm wildlife.
3) For a farm to be classed as organic, it will usually have to follow guidelines on the <u>ethical treatment of animals</u>. This means <u>no</u> battery farming.
4) Many people feel that the <u>quality</u> of organic food is better, e.g. the flavour.

DISADVANTAGES

1) Organic farming takes up <u>more space</u> than intensive farming — so more land has to be <u>farmland</u>, rather than being set aside for wildlife or for other uses.
2) It's more <u>labour-intensive</u>. This provides <u>more jobs</u>, but it also makes the food more <u>expensive</u>.
3) You can't grow <u>as much</u> food as you can with intensive farming.

You Can Investigate the Best Conditions for Plant Growth

To investigate how well <u>plants grow</u>, stick some <u>seeds</u> in some pots, <u>vary the conditions</u> and see which ones <u>grow</u> the best. Sounds easy — but here are some things you've got to consider:

1) What <u>FACTOR</u> are you investigating?

   This could be <u>temperature</u>, amount of <u>water</u>, amount of <u>light</u>, type of <u>fertiliser</u>, or presence of a <u>herbicide</u>. You've got to make sure that <u>every other factor</u> remains the <u>same</u> throughout the experiment — this is usually the hard bit.

2) How are you going to <u>MEASURE</u> the growth?

   You could count the <u>number of leaves</u>, or measure the plant's <u>height</u> or <u>mass</u>. (Or, if you're anything like me, which one stays alive the longest.)

You should <u>repeat</u> the experiment a couple of times, to make sure you get the same results — this makes your results more <u>reliable</u>.

There's nowt wrong wi' spreadin' muck on it...

You may well have quite a strong opinion on stuff like intensive farming of animals — whether it's 'tree-hugging hippie liberals, just give me a bit of nice cheap pork,' or 'poor creatures, they should be free, free as the wind!' Either way, keep it to yourself and give a nice, balanced argument in the exam.

Revision Summary for Section 2.3

Hmm... so this whole farming lark, what's that all about? Well, if you don't know, you haven't read this section properly, and your first task is to go back and read it all again. However, if you think you know your intensive farming methods from your organic farming methods, here are some questions for you.

1) Why do plants need nitrates?

2) What could happen to a plant if there wasn't enough potassium in the soil?

3) What is magnesium needed for? What might happen to plants without enough magnesium?

4) Why do farmers need to add minerals to the soil?

5) What is intensive farming?

6) Why does intensive farming use fertilisers?

7) What is a chemical that's used to kill weeds called?

8) What is battery farming and why is it used?

9) State one alternative to intensive farming.

10) How does this method keep the land fertile?

11) How does the treatment of animals differ between this method and intensive farming?

12) What is meant by 'biological control'?

13) Give two advantages of intensive farming.

14) How does removing hedges have a bad effect on the environment?

15) Give two advantages and two disadvantages of organic farming.

16) Plan an experiment to find out if you get better plant growth using compost or soil from a garden. Give three factors you'll need to keep constant. How would you measure the growth?

Avoiding Contamination

Forensic science is about <u>using science</u> to <u>help solve crimes</u>. It's used in all types of crime, e.g. murder, rape, drugs, arson, terrorism, forgery, fraud and burglary.

Forensic Science Involves Two Main Stages

A forensic scientist's job is to <u>help</u> the <u>police</u> figure out what has happened and who has committed the crime. They do this by looking at the <u>physical evidence</u>. There are two main stages:

1) The evidence is <u>collected</u> and <u>recorded</u> — this section is all about evidence collection.

2) The evidence is <u>analysed</u> — once you've got the evidence, you need to figure out <u>what it means</u>. This could involve chemically analysing a white powder found at the crime scene to figure out what it is, or comparing a fingerprint found at the crime scene to a suspect's prints.

The findings of the analysis are given to the investigating <u>police officer</u>.
The forensic scientist might also be asked to <u>present their findings</u> in <u>court</u>.

At a Crime Scene You Need to Avoid Contamination

Contamination is where something is added that <u>wasn't originally at the crime scene</u> (e.g. hair or blood). It's really important to avoid contamination because it makes it <u>harder</u> for the police to figure out <u>what happened</u> and it also might make the <u>evidence invalid</u> (so it can't be used in court).
There are <u>three</u> main ways to avoid contamination:

1) RESTRICTING ACCESS

1) The <u>first police officer</u> at the crime scene will make sure that the crime scene is <u>safe</u>, i.e. any suspects have been arrested or the crime scene is <u>empty</u>.

2) They will then <u>stop</u> any <u>unauthorised people</u> from <u>entering</u> the crime scene. They may also <u>cover up</u> any evidence that might be blown away or washed away by rain. This <u>prevents</u> the evidence from being <u>disturbed</u> and stops any <u>'foreign' material</u> being <u>introduced</u> into the crime scene. It also stops someone <u>planting evidence</u> or <u>removing vital evidence</u>.

2) WEARING PROTECTIVE CLOTHING

1) The <u>forensic scientists</u> who specialise in collecting evidence from crime scenes will then arrive.

2) They wear <u>protective clothing</u> to prevent <u>contaminating the scene</u> with anything that's on themselves or on their shoes, e.g. hairs or mud. This also <u>protects</u> the scientists from <u>potentially harmful substances</u> at the scene, e.g. blood and chemicals.

face mask
shoe covers
full protective suit to avoid leaving hairs etc.
plastic gloves to avoid leaving fingerprints

MICHAEL DONNE / SCIENCE PHOTO LIBRARY

3) Using the right COLLECTION, STORAGE and RECORDING METHODS

1) The scientists need to collect all the evidence <u>separately</u> to stop <u>one bit</u> of evidence <u>contaminating another</u>.

There's more about this on the next page.

2) This means collecting <u>each piece</u> of evidence with a clean instrument, e.g. using one swab for each different sample of blood collected, and <u>storing</u> each piece in a <u>different bag</u> or <u>container</u>.

Avoid contamination — don't put red socks in with your whites...

It's pretty important to avoid contamination. E.g. if the suspect somehow got back into the crime scene then any evidence linking them to the crime might be <u>invalid</u>. The Police <u>can't be sure</u> what was <u>original evidence</u> and what was <u>introduced</u> by the suspect re-entering the scene. It's a defence lawyer's dream...

Collecting Evidence

Once the crime scene is <u>secure</u> the forensic scientists can start to <u>collect</u> and <u>record</u> the evidence.

Different Things are Collected by Different Methods

<u>How</u> you collect a piece of evidence and what you <u>store it in</u> depends on <u>what the evidence is</u>, e.g. you're not going to store a wet T-shirt in a paper bag.

Here are a few examples of how different pieces of evidence are collected and stored:

1) FIBRES — These are often collected using <u>sticky tape</u>. The tape is then <u>stuck to something</u>, e.g. a piece of plastic, and <u>sealed</u> in a <u>plastic bag</u>. That way individual fibres <u>don't get lost</u>.

2) SOIL — A <u>small sample</u> is usually collected and put in a <u>plastic pot</u>. This may then be <u>sealed</u> in a <u>plastic bag</u> just in case it <u>spills</u>.

3) FINGERPRINTS — These are collected by <u>dusting</u> and <u>lifting</u> them off with sticky tape or by <u>photographing them</u> (see p.40 for more).

4) BROKEN GLASS — Pieces of glass are collected and placed in a <u>paper bag</u>.

5) BLOOD — <u>Small samples</u> of blood are collected using a <u>swab</u> (a bit like a cotton bud) and <u>sealed</u> in a plastic tube or plastic bag. Things that are <u>wet</u> with blood are <u>sealed</u> in a <u>plastic bag</u> and often <u>frozen</u> to stop them going mouldy.

How Much is Collected Depends on What It Is

If you've got <u>loads</u> of one type of evidence you only need to collect a <u>small sample</u>. You just need <u>enough</u> to <u>run</u> all your <u>tests on</u>. Here are some examples:

1) If you need a <u>hair sample</u> from a victim or suspect, you only need to take a <u>small snip</u> of hair (or pull a few hairs out), not make them bald.

2) If a <u>pond</u> has been polluted by a nearby factory you only need to take a <u>couple of hundred millilitres</u> for testing, not all the water in the pond.

Everything Has to be Recorded and Labelled

1) As the evidence is collected a <u>detailed record</u> has to be made and the evidence has to be <u>labelled</u>.

2) The <u>record</u> includes the following things:
   - The <u>time</u> and <u>date</u> the evidence was collected.
   - <u>Where</u> the evidence was <u>found</u> in the <u>crime scene</u>.
   - <u>Who</u> collected it.
   - A <u>brief description</u> of what it is.

3) The bag containing each piece of evidence is also labelled with a <u>reference number</u> (which matches the record), a <u>brief description</u> and other police and court details. Every person who examines the evidence has to <u>sign and date</u> the back of the label so they can <u>keep track</u> of who has had it.

If you don't collect, record and label all the evidence <u>properly</u> then the evidence might be deemed <u>unreliable</u> and <u>can't be used</u> as <u>evidence in court</u>. Here's an example:

A victim was stabbed and Police find a man nearby whose shirt is splattered with blood (which they think is from the victim). If the victim's T-shirt is collected and stored with the suspect's T-shirt you <u>can't be sure</u> whether the suspect's T-shirt originally had the victim's blood on it or not.

Hide the evidence — don't leave the sweet wrappers behind...

In the Exam you might be asked to suggest <u>how much</u> of the evidence to collect. Just use your <u>common sense</u> — if there's loads of it take a little and if there's hardly any of it take it all.

Recording Impressions and Marks

In forensics an impression isn't what you make on your girlfriend's parents, it's an <u>indent</u> left in something by another object. Impressions and marks can tell scientists who and what has been where.

Impressions are Indents Left in a Substance

Impressions can be <u>made by</u> loads of things, e.g. tyres, shoes, hands, tools and even teeth. They can be <u>left in</u> different substances, e.g. mud, soil, dust, metal, wood, food and bodies.

There are Different Methods to Record Impressions

1) Forensic scientists need a <u>permanent record</u> of the impression so they can analyse it.

2) It's also important to make a permanent record because some impressions can <u>disappear quickly</u> from a crime scene (e.g. they can be washed away).

3) An impression is recorded by <u>making a cast</u> (mould) of it.

4) <u>Before</u> a cast of any kind is made the impression <u>must be PHOTOGRAPHED</u>. This means there's a permanent record, <u>just in case</u> the scientists stuff up making the cast or there's a torrential downpour while they're doing it and the whole thing washes away.

5) The method used to make the cast depends on <u>what substance</u> the impression is left in:

1) USING PLASTER OF PARIS

— This is a white <u>powder</u> that is <u>mixed</u> with <u>water</u> to produce a <u>liquid</u>. The liquid is <u>poured</u> into the impression and allowed to <u>set</u>, producing a <u>solid plaster cast</u> of the impression. This method is used to record impressions in <u>soft substances</u>, e.g. footprints in mud or tyre prints in soil.

2) USING PLASTICINE

— This can be <u>pushed</u> into an impression to form a <u>plasticine cast</u> of it. This method could be used to record impressions from <u>solid objects</u>, e.g. a screwdriver mark in a car door.

Marks Can be Many Things

Marks are pretty much anything you can see that isn't an impression, e.g. pen marks, scratches, spots of blood, splodges, smears of paint...

Marks are Recorded by Taking a Photograph

Marks are <u>recorded</u> by <u>taking a photograph</u> of them. A <u>scale</u> is included in the photograph so scientists know the <u>size</u> of the mark.

I make a great first impression...

Don't forget impressions need to be <u>photographed first</u>, just in case you muck up making the cast by removing it before it has set properly and it all falls apart. In real-life forensic work they don't tend to use plasticine (because it might get squished) — they often use <u>rubber</u> to make a cast of impressions.

Comparing Impressions and Marks

An impression by itself <u>isn't very useful</u> — it won't tell you <u>whose</u> screwdriver, shoe, tyre (or whatever) made it. You need to <u>compare</u> it with an impression made by <u>another object</u> — one owned by the suspect.

Impressions are Then Compared with Real Objects

Different kinds of object leave <u>different marks</u> that forensic scientists use to compare two samples. Here are a couple of examples of what they look for:

Tyre Marks

1) <u>Width</u> of tyre mark.

2) <u>Tread pattern</u> — every <u>make</u> of tyre has a different tread pattern (this is the pattern of <u>raised ridges</u> on the surface of the tyre).

Shoe Marks

1) <u>Width</u> of shoe.

2) <u>Length</u> of shoe.

3) <u>Tread pattern</u> — most makes of shoes have different tread patterns.

Tool Marks

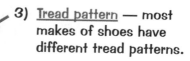

1) <u>Width</u> of mark.

2) <u>Length</u> of mark.

3) <u>Shape</u> of mark, e.g. a saw will leave a different shaped mark to a crowbar or a hammer.

These things may tell you if your two samples are the <u>same make</u> but they <u>won't</u> tell you if they're <u>exactly the same object</u>. To tell if they're exactly the same object you have to look for <u>matching distinctive marks</u> (marks that are <u>unique</u> to that object), e.g. a crack on a tyre, a pin stuck in a shoe or a saw tooth missing.

Here's an Example

1) A fresh <u>footprint</u> is found in the <u>sand</u> next to a factory that has been burned down.

2) Forensic scientists <u>collect</u> the impression using <u>plaster of Paris</u>.

3) The police suspect it <u>may</u> have been a <u>disgruntled employee</u> that set it alight. They collect all the <u>employees' shoes</u> and the scientists make <u>comparison shoe prints</u>. Here are some of the results:

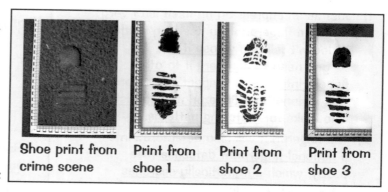

Shoe print from crime scene | Print from shoe 1 | Print from shoe 2 | Print from shoe 3

The <u>length</u> of shoe 1 is <u>too long</u> so it can't have been that one, the <u>tread pattern</u> on shoe 2 is <u>different</u> so it can't have been that one either. The length and tread pattern of shoe 3 <u>match</u> the sample from the crime scene so the employee wearing shoe 3 <u>might</u> have been at the crime scene.

Learn this page and you'll get good marks...

In the exam you could be asked to <u>suggest</u> which <u>measurements</u> or distinctive marks could be used to compare marks or impressions. It's pretty easy really — go for <u>size</u>, <u>width</u>, <u>depth</u> and <u>patterns</u>.

Taking Fingerprints

The <u>ridges of skin</u> on your fingertips have <u>distinctive patterns</u>. When you touch something you leave a <u>print</u> of the pattern (in sweat... yuk) on the surface you touched. This is called a <u>fingerprint</u>.

There are Two Main Methods for Collecting Fingerprints

How you collect a fingerprint from a <u>crime scene</u> depends on whether the fingerprint is <u>visible</u> or <u>not</u>.

1) PHOTOGRAPHING IT — If the person has a <u>dark coloured substance</u> like blood or ink on their fingers they will make a <u>visible fingerprint</u>. The fingerprint is recorded by <u>taking a photograph</u> of it.

2) DUSTING AND LIFTING WITH TAPE

— If you <u>can't see the fingerprint</u> the scientist uses a <u>brush</u> to apply a <u>dark coloured dust</u>. The dust <u>sticks</u> to the fingerprint, making it <u>visible</u>. A <u>photograph</u> of the fingerprint is then taken (just in case they mess up the next bit). A piece of <u>sticky tape</u> is then stuck to the fingerprint and when it is peeled off a visible copy of the fingerprint is made. Sometimes forensic scientists use a hi-tech method of lifting the print called <u>electrostatic dust-lifting</u>.

Fingerprints Have Three Types of Pattern

1) The three main <u>patterns</u> on your fingertips are called <u>loops</u>, <u>arches</u> and <u>whorls</u>.

2) Each fingerprint can have <u>one</u> pattern or a <u>combination</u> of the three types.

Whorl   Arch   Loop

3) The type of pattern and the location of the pattern on the fingertip are <u>different for every person</u> (even identical twins have different fingerprints), which means that fingerprints are <u>unique</u>.

4) Each of your <u>fingers</u> also has a <u>different fingerprint</u>.

The Fingerprint is Compared with Those of the Suspects

1) Just as an impression by itself isn't very useful, a fingerprint by itself isn't either. It doesn't tell you <u>whose it is</u> — you need to <u>compare</u> it to <u>other fingerprints</u>, e.g. a suspect's fingerprints. This is done by a <u>fingerprint specialist</u>, who looks for <u>matching patterns</u>.

2) The fingerprint will also be entered into a <u>national fingerprint database</u> (see p.59), which <u>automatically searches</u> for a <u>possible match</u>.

Here's an example:

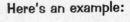

Fingerprint from crime scene   Suspect 1's fingerprint   Suspect 2's fingerprint

The loop on the fingerprint from the crime scene matches the loop on Suspect 2's fingerprint.

Fingerprint specialists have a whorl of a time...

Prints of other body parts can also be used by forensic scientists in the same way, e.g. <u>palm</u> prints and <u>sole</u> prints have distinctive skin patterns, which can be used to narrow down identities. Forensic scientists can even nab burglars by <u>ear</u> prints they leave when listening at windows...

Revision Summary for Section 2.4

That was a pretty fascinating section I reckon... I never knew how many different types of evidence forensic scientists looked for. It makes you think just how much evidence you leave about for them to find — just on your way to school you'll leave footprints, bits of hair, fingerprints all over doors etc. Anyway, you need to know this stuff for your exam... so have a go at the questions below to check what you know and what you don't. If there's one you can't answer, go back to the page and have another look.

1) What is contamination?
2) Why is it important to avoid contamination at a crime scene?
3) Why do forensic scientists wear protective clothing at a crime scene?
4) Name three items of protective clothing a forensic scientist might wear at a crime scene.
5) Why is it important to collect and store evidence separately?
6) Describe how you would collect fibres from a crime scene.
7) Describe how you would collect a small sample of blood from a crime scene.
8) Describe how you would collect a sample of soil from a crime scene.
9) What are the five things that must be recorded for every piece of evidence collected from a crime scene?
10) What might happen if the evidence is not collected, recorded and stored properly?
11) Name three objects that can make impressions.
12) Describe two different methods of recording impressions.
13) What does the forensic scientist do before they make a cast of an impression?
14) How are marks normally recorded?
15) What must you include in the scene when photographing evidence?
16)* What method would you use to record a hand print made in mud?
17) What do forensic scientists do once they have recorded an impression?
18) What distinctive marks would you look for when comparing shoe prints?
19)* What distinctive marks might you look for when comparing two hand prints?
20) What is a fingerprint?
21) How would you collect a fingerprint made in ink?
22) Describe another method of collecting fingerprints.
23) Draw a diagram of the three different types of fingerprint pattern.
24) Does anyone have the same fingerprint as anyone else?
25)*What will happen to a fingerprint found at a crime scene if there is no suspect for the crime?

* Answers on page 100.

Ionic Compounds

Once the evidence has made it from the crime scene back to the lab, it's time to <u>analyse</u> it, and that's what this section is all about — <u>chemical analysis</u>.

The <u>Appearance of a Substance</u> Gives the <u>First Clues</u>

The first thing you're likely to notice about a substance is its <u>appearance</u> — including its colour and whether it's a solid, a liquid or a gas. However, a suspicious bag of a <u>white powder</u> could be loads of different substances, from <u>drugs</u> to <u>talcum powder</u>.

The <u>behaviour</u> of a compound gives us the next clues — this is determined by the <u>structure</u> and <u>bonding</u> within the substance.

Ionic Compounds Have a <u>Giant Lattice</u> Structure

Many solids you'll encounter in your day-to-day life are <u>ionic</u>.

1) Ionic compounds are made of <u>charged particles</u> called <u>ions</u>.
2) <u>Metal ions</u> are always <u>positively</u> charged.
3) <u>Non-metal ions</u> are usually <u>negatively</u> charged.
4) Ions with opposite charges are <u>strongly attracted</u> to one another. You get a massive <u>giant lattice</u> of ions built up — like the one on the right here...
5) This is the normal kind of bonding between a <u>metal</u> and <u>non-metal</u> — called <u>ionic bonding</u>.

A common example is <u>sodium chloride</u> (table salt). It has <u>positive sodium ions</u> (the metal ions) and <u>negative chloride ions</u> (the non-metal ions).

Ionic Compounds Have <u>High Melting Points</u>

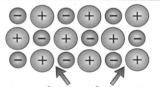

strong forces of attraction

1) The <u>forces of attraction</u> between the ions are <u>very strong</u>.
2) It takes a lot of <u>energy</u> to overcome these forces and <u>melt</u> the compound.
3) So ionic compounds have <u>high melting points</u>.
4) Which makes them <u>solids</u> at room temperature.

Some Ionic Compounds <u>Dissolve</u> in Water

Some ionic compounds are <u>soluble in water</u>, but some aren't. This makes life a little more difficult for you, but a little easier for forensic scientists when they're trying to work out which ionic compound they have. Here are some really handy rules:

1) All <u>sodium</u> (Na), <u>potassium</u> (K) and <u>ammonium</u> (NH_4) salts are <u>SOLUBLE</u> in water.
2) All <u>nitrates</u> (NO_3) are <u>SOLUBLE</u> in water.
3) Most <u>chlorides</u> (Cl) are <u>SOLUBLE</u> in water, except for silver and lead.
4) Most <u>sulfates</u> are <u>SOLUBLE</u> in water, except for barium and lead. Calcium sulfate is <u>slightly soluble</u>.
5) Most <u>oxides</u> (O), <u>hydroxides</u> (OH) and <u>carbonates</u> (CO_3) are <u>INSOLUBLE</u> in water, except for sodium and potassium.

<u>Giant ionic lattices — all over your chips...</u>

To identify which <u>metal ion</u> is present you could use the flame test on p.45 or the precipitation test with sodium hydroxide on p.47. The <u>non-metal ions</u> could be carbonates (p.46), sulfates (p.47) or chlorides (p.47). You can even test for its solubility using the method on p.46. All this yet to come...

Formulas of Ionic Compounds

Once the positive and negative ions have been identified you can work out the formula. Lucky you.

The Charges in an Ionic Compound Add Up to Zero

Different ions have <u>different charges</u>, shown in the table:

Some metals (like iron, copper and tin) can form ions with <u>different charges</u>. The number <u>in brackets</u> after the name tells you the <u>size</u> of the <u>positive charge</u> on the ion — and luckily for us, this makes the charge really easy to remember. E.g. an iron(II) ion has a charge of 2+, so it's Fe^{2+}.

The main thing to remember is that in compounds the <u>total charge must always add up to zero</u>.

Positive Ions		Negative Ions	
Lithium	Li^+	Chloride	Cl^-
Sodium	Na^+	Cyanide	CN^-
Potassium	K^+	Oxide	O^{2-}
Magnesium	Mg^{2+}	Carbonate	CO_3^{2-}
Calcium	Ca^{2+}	Sulfate	SO_4^{2-}
Iron(II)	Fe^{2+}		
Iron(III)	Fe^{3+}		
Aluminium	Al^{3+}		

The Easy Ones

If the ions in the compound have the <u>same size charge</u> then it's easy.

> **EXAMPLE:** Find the formula for <u>calcium oxide</u>.
>
> Find the charges on a calcium ion and an oxide ion.
> A calcium ion is Ca^{2+} and an oxide ion is O^{2-}.
> To balance the total charge you need one calcium ion to every one oxide ion.
> So the formula of calcium oxide must be: **CaO**

> **EXAMPLE:** Find the formula for <u>potassium cyanide</u>.
>
> Find the charges on a potassium ion and a cyanide ion.
> A potassium ion is K^+ and a cyanide ion is CN^-.
> To balance the total charge you need one potassium ion to every one cyanide ion.
> So the formula of potassium cyanide must be: **KCN**

The Slightly Harder Ones

If the ions have different size charges, you need to put in some numbers to balance things up.

> **EXAMPLE:** Find the formula for <u>sodium oxide</u>.
>
> Find the charges on a sodium ion and an oxide ion.
> A sodium ion is Na^+ and an oxide ion is O^{2-}.
> To balance the total charge you need two sodium ions to every one oxide ion.
> So the formula of sodium oxide must be: **Na_2O**

> **EXAMPLE:** Find the formula for <u>aluminium chloride</u>.
>
> Find the charges on an aluminium ion and a chloride ion.
> An aluminium ion is Al^{3+} and a chloride ion is Cl^-.
> To balance the total charge you need one aluminium ion to every three chloride ions.
> So the formula of aluminium chloride must be: **$AlCl_3$**

The Dead Hard Ones

You'll need to do a fair bit of balancing to get these ones sorted.

> **EXAMPLE:** Find the formula for <u>aluminium oxide</u>.
>
> Find the charges on an aluminium ion and an oxide ion.
> An aluminium ion is Al^{3+} and an oxide ion is O^{2-}.
> To balance the total charge you need two aluminium ions to every three oxide ions.
> So the formula of aluminium oxide must be: **Al_2O_3**

> **EXAMPLE:** Find the formula for <u>iron(III) sulfate</u>.
>
> Find the charges on an iron(III) ion and a sulfate ion.
> An iron(III) ion is Fe^{3+} and a sulfate ion is SO_4^{2-}.
> To balance the total charge you need two iron(III) ions to every three sulfate ions. So the formula of iron(III) sulfate must be: **$Fe_2(SO_4)_3$**

Any old ion, any old ion — any, any, any old ion...

After all those examples, I'm sure you could work out the formula to any ionic compound. And just to test that theory here are a few for you to try: a) magnesium oxide, b) lithium oxide, c) sodium sulfate.*

*Answers on p.100.

Covalent Compounds

Forensic scientists could also come across <u>covalent compounds</u>. These substances may well come from living materials. So they could be dealing with blood, guts and gore. Ughh!

Many <u>Covalent Substances</u> Come from <u>Living Materials</u>

Many substances obtained from <u>living materials</u> are <u>organic compounds</u>.

1) Organic compounds tend to have <u>covalent bonding</u> (see below).

2) Covalent substances usually contain <u>non-metals</u>.

Forensic scientists analyse covalent compounds in <u>fluids</u>, like blood and urine, to help them get a better picture of what they're investigating.

Organic covalent compounds, e.g. ethanol, glucose, drugs

1) <u>Ethanol</u> (C_2H_5OH) — The <u>concentration</u> of ethanol (alcohol) in the <u>blood</u> is measured to find out if someone's '<u>over the limit</u>' whilst driving (see p.48).

2) <u>Glucose</u> ($C_6H_{12}O_6$) — The <u>concentration</u> of glucose in the <u>blood</u> and in <u>urine</u> (see p.48) can indicate if a person is <u>diabetic</u>.

3) <u>Methanol</u> — Methanol is a <u>poisonous alcohol</u> that has been found in bottles of vodka, cheap wine and other alcoholic drinks from illegal distilleries.

4) <u>Drugs</u> — Blood analysis can show if a person's taken <u>illegal drugs</u>, e.g. ecstasy or cocaine.

Inorganic covalent compounds, e.g. water, carbon dioxide

1) <u>Water</u> (H_2O) — If water's found in the lungs of a body it suggests that they've died from <u>drowning</u>.

2) <u>Carbon dioxide</u> (CO_2) — An increased concentration of carbon dioxide in the blood of a dead person may indicate that they have <u>suffocated</u> — from choking, drowning, or inhalation of toxic gases.

Covalent Compounds Have <u>Low Melting and Boiling Points</u>

1) Covalent compounds usually exist as <u>small molecules</u>.

2) The atoms within the molecules are held together by strong forces called <u>covalent bonds</u>.

3) In contrast, the <u>forces of attraction between</u> these molecules are <u>very weak</u>.

4) You only need a little bit of energy to overcome the weak forces between the molecules — so covalent compounds have <u>low melting points</u> and <u>boiling points</u>.

5) And this means they're usually <u>gases</u> and <u>liquids</u> at room temperature.

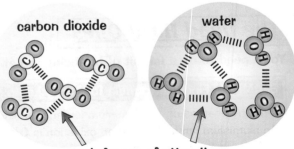

carbon dioxide

water

weak forces of attraction

<u>H_2O, CO_2, C_2H_5OH, DVD, FBI, GSOH...</u>

Okay. So many substances obtained from living materials are <u>organic</u>. And organic compounds have <u>covalent bonding</u>. Got that? Now have a good look at the formulas of the covalent compounds — they'll expect you to know these. Cover up the page, scribble them down and check you've got them right.

Flame Tests and pH

Imagine the police are investigating a murder, and they've found a curious <u>white powder</u> near the blood-stained candlestick. It's time to start analysing the powder, with the hopes of finding some more clues...

Flame Tests Test for Metal Ions

<u>Flame tests</u> can identify which <u>metals</u>, if any, are present in a compound. If you put even <u>minute</u> amounts of some metals into a flame, very distinctive <u>colours</u> can be seen.

> To do a flame test you need a <u>nichrome wire loop</u>, dilute <u>hydrochloric acid</u> (HCl) and a blue <u>Bunsen flame</u>.
>
> 1) First make sure the nichrome wire loop is <u>really clean</u>, or you might <u>contaminate</u> your sample. Usually it's alright to dip the loop into <u>hydrochloric acid</u> and then rinse it in <u>distilled water</u>, but if it's really dirty you might have to use <u>emery cloth</u>.
> 2) <u>Dip</u> the wire in your sample (you only need a little).
> 3) Put the wire loop in the <u>blue</u> part of the <u>Bunsen flame</u> (the hottest bit).
> 4) <u>Observe</u> and <u>record</u> the results.

Compounds of some metals burn with a <u>characteristic colour</u>, as you see every November 5th. So, remember, remember...

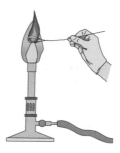

> 1) <u>Sodium</u>, Na^+, burns with an orange flame.
> 2) <u>Potassium</u>, K^+, burns with a lilac flame.
> 3) <u>Calcium</u>, Ca^{2+}, burns with a brick-red flame.
> 4) <u>Copper</u>, Cu^{2+}, burns with a blue-green flame.

Universal Indicator Paper Measures pH

An indicator is just a dye that <u>changes colour</u> depending on whether it's <u>in an acid</u> or <u>in an alkali</u>. <u>Universal indicator</u> is a very useful <u>combination of dyes</u> which gives the colours shown below.

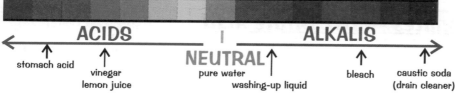

Universal indicator produces a range of different colours.

> The pH scale goes from 0 to 14
> 1) A <u>very strong acid</u> has <u>pH 0</u>. A <u>very strong alkali</u> has <u>pH 14</u>.
> 2) If something is <u>neutral</u> it has <u>pH 7</u> (e.g. pure water).
> 3) Anything <u>less</u> than 7 is <u>acid</u>. Anything <u>more</u> than 7 is <u>alkaline</u>.

To measure the pH of a substance all you need to do is pipette a <u>drop</u> of it onto <u>universal indicator paper</u>, and record its <u>colour</u>. If it's a <u>solid sample</u> then you'll need to make it into a <u>solution</u> first (see p.46).

Example: Scientists are able to distinguish <u>cocaine</u> from a lot of other powders as it has a <u>high pH</u>.

An untidy bedroom — the universal indicator for a teenager...

A more accurate method of measuring pH is to use a <u>pH meter</u> — it tells you the pH digitally so there's no messing about with trying to work out which colour the universal indicator paper resembles the most.

Solubility and Carbonates

With a little bit of this, and a little bit of that... you'll soon work out the identity of that mystery substance.

Finding the Solubility of a Sample

By finding out the solubility of a compound you can eliminate a lot of ionic compounds (see p.42).

And here's how to do it:

1) Add a very small amount of the substance to some distilled water in a boiling tube and shake.

2) After allowing it to settle, record what the contents look like.

3) If it's clear, the substance is soluble. If it's cloudy, it's slightly soluble, and if there's no change, it's insoluble.

SHAKE

Clear   Cloudy   No Change

You've got to be very careful with this test — if you put in too much, the solution may become saturated, so it'll look like the soluble substance is insoluble. A saturated solution is one that will not dissolve any more of the solid (unless you change the temperature).

Making a Soluble Sample into a Solution

For many chemical tests the substance needs to be in solution, like testing for pH (see p.45), and chromatography (see p.58).

Here's the method:

1) Add a spatula of the sample to some distilled water in a boiling tube and shake.

2) Repeat this process until no more solid will dissolve.

3) Pour the solution through some filter paper to remove any excess solid.

4) Store the solution for future testing.

Testing for Carbonates — Use Dilute Acid

Most carbonates are insoluble (see p.42). They're in many everyday items, like sodium carbonate, found in detergents, and calcium carbonate, which is found in cement and in antacids.

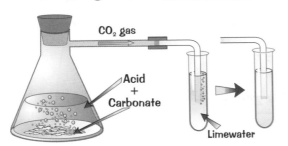

CO₂ gas

Acid + Carbonate

Limewater

Carbonates give off carbon dioxide when added to dilute acids. Here's the method:

1) Put your mystery compound in dilute acid, e.g. dilute hydrochloric acid, and collect any gas given off.

2) Bubble the gas through limewater. If the limewater turns milky, the gas given off is carbon dioxide...

3) ...so your compound contains carbonate ions — CO_3^{2-}.

If you're not part of the solution, you're part of the precipitate...

Illegal drugs, like heroin and cocaine, can be 'cut' or mixed with other white powders to increase their street value. These can include washing soda, which contains a carbonate. Other common ones are sugar, talcum powder and baby milk powder — it's amazing what stuff people will snort up their noses...

Precipitation Tests

In precipitation tests, two dissolved substances react to form an <u>insoluble</u> solid — called the <u>precipitate</u>. These can help you identify soluble ionic compounds.

Testing for <u>Metal Ions</u> — <u>Sodium Hydroxide</u>

1) Many <u>metal hydroxides</u> are <u>insoluble</u> — so they precipitate out of solution when formed.

2) Some of these hydroxides have a <u>characteristic colour</u>.

3) In this test you just add a few drops of <u>sodium hydroxide</u> solution to your mystery solution, and see what happens.

4) If a precipitate forms, it's colour can tell you which <u>metal hydroxide</u> you've made — and so what the <u>metal bit</u> of your mystery compound could be...

Metal Ion	Colour of Precipitate
Calcium, Ca^{2+}	White
Copper, Cu^{2+}	Blue
Iron(II), Fe^{2+}	Sludgy Green
Iron(III), Fe^{3+}	Reddish Brown
Lead(II), Pb^{2+}	White at first. But if you add loads more NaOH it forms a colourless solution.

E.g. calcium chloride + sodium hydroxide → calcium hydroxide + sodium chloride

Testing for <u>Sulfates</u> — <u>Hydrochloric Acid</u> then <u>Barium Chloride</u>

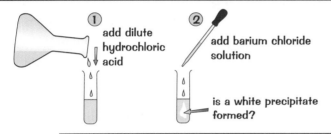

add dilute hydrochloric acid

add barium chloride solution

is a white precipitate formed?

1) Add some <u>dilute hydrochloric acid</u> to a solution of your compound.

2) Then add a few drops of <u>barium chloride solution</u> to the liquid.

3) If you see a <u>white precipitate</u>, there are <u>sulfate</u> ions (SO_4^{2-}) in your compound.

E.g. barium chloride + copper sulfate → barium sulfate + copper chloride

Testing for <u>Chlorides</u> — <u>Nitric Acid</u> then <u>Silver Nitrate</u>

1) Add <u>dilute nitric acid</u> to a solution of your compound.

2) Then add a few drops of <u>silver nitrate solution</u> to the liquid.

3) If you see a <u>white precipitate</u>, there are <u>chloride</u> ions (Cl^-) in your compound.

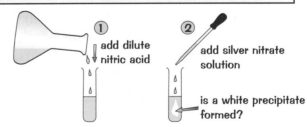

add dilute nitric acid

add silver nitrate solution

is a white precipitate formed?

E.g. silver nitrate + copper chloride → silver chloride + copper nitrate

You Can Test <u>Water</u> for Various <u>Dissolved Ions</u>

Many ions, like nitrates and lead, are <u>poisonous</u> and it can be dangerous if they get into the water supply. There are <u>strict legal limits</u> on the levels of these ions in rivers — so companies who put their <u>waste</u> in rivers have to make sure they don't exceed them.

1) Forensic scientists take <u>samples</u> from <u>rivers</u> to make sure <u>pollutant levels</u> don't <u>exceed</u> these limits.

2) If poisonous ions are found in the water, <u>further investigation</u> will be carried out to find the <u>culprit</u>.

Snow White Precipitate — and the Seven Analytical Chemists...

The coloured compound in each of the equations above shows the <u>insoluble product</u> of the reaction. You may be asked to name these — just swap over the first and second bits of the names of the reactants.

Tests for Ethanol and Glucose

These next two tests involve organic covalent compounds. If you've forgotten what they are, have a quick flick back to p.44. I'm sure you'll be glad to hear that they're the last tests in this section. Enjoy.

Use Acidified Potassium Dichromate to Test for Ethanol

It's illegal in the UK to 'drink and drive'. The amount of alcohol is measured using a breathalyser — which uses a chemical-based test to detect the presence of ethanol on a suspect's breath.

1) Acidified potassium dichromate solution is orange.
2) When ethanol is added the solution goes green.
3) This provides a simple colour change test to find out if the sample contains ethanol.

orange → green

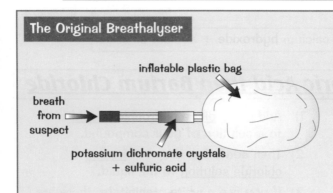

The Original Breathalyser

inflatable plastic bag

breath from suspect

potassium dichromate crystals + sulfuric acid

1) The suspect is asked to blow into the tube until the inflatable bag is full.
2) Potassium dichromate crystals are contained in a small chamber, mixed with sulfuric acid.
3) Any ethanol found in their breath will react with the potassium dichromate and turn it green.
4) The police officer then compares the colour with an unreacted sample — if the colour change is significant the suspect has failed the test.

Modern breathalysers use a similar test, but they give you a digital read-out of the amount of alcohol in your breath. Unfortunately, even modern breathalysers aren't very accurate, and if the result is positive the suspect has to be taken into the police station where more accurate tests can be carried out.

Use the Benedict's Test for Glucose

1) You can test for glucose in an aqueous (water-based) solution such as blood or urine by using the Benedict's test.
2) In a boiling tube add about 3 cm³ of the solution you are testing to the same volume of Benedict's solution.
3) Gently warm the boiling tube in a hot water bath over a Bunsen flame.
4) A coloured precipitate gradually forms if glucose is present. The colour change goes like this:

   blue → green → yellow → orange → brick red

5) The higher the concentration of glucose, the further the colour change goes.
6) You can also do the Benedict's test on a solid sample — add 3 cm³ of Benedict's solution into a boiling tube with a spatula of the sample and do the test in the same way.

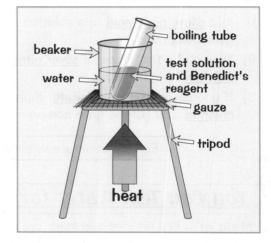

beaker

water

boiling tube

test solution and Benedict's reagent

gauze

tripod

heat

Ethanol and glucose — a couple of my favourite things...

Nowadays you don't get to breathe into a plastic bag — sad isn't it. Instead you have to blow into a boring looking box which digitally tells you if the result is positive or negative. Once inside the police station they'll ask for a blood or urine sample so they can measure the alcohol concentration directly.

Identifying a Compound

A single chemical test isn't likely to identify a compound, so forensic scientists usually carry out a series of tests and the evidence from all of them is used to draw conclusions.

A Range of Different Tests Will Usually be Carried Out

EXAMPLE: A notorious <u>drug dealer</u> has been caught in possession of a large quantity of an unknown white powder, which is later identified as <u>cocaine</u>. Investigators would like to know what he used to <u>cut</u> the drug with, so that they can trace other drug samples back to the dealer.

They raid his house and find four different <u>household</u> white powders shown in the table.

Investigators are given a <u>sample</u> of the powder and are asked to find out which <u>household substance</u>, if any, is present. Their <u>final report</u> is shown below.

Substance	Chemical Name	Chemical Formula
table salt	sodium chloride	NaCl
washing powder	sodium carbonate	Na_2CO_3
corn sugar	glucose	$C_6H_{12}O_6$
chalk	calcium carbonate	$CaCO_3$

Forensic Laboratory Report

PRIORITY: Urgent

DETAILS: White powder recovered from suspect. Check for possible match with substances retrieved from suspect's residence.

TEST	OBSERVATION	INFERENCE
Solubility	Does dissolve	Sample is soluble in water.
pH	pH 9	Sample is slightly alkaline.
Benedict's Test	No colour change	Sample doesn't contain glucose.
Flame Test	Orange flame	Sample contains sodium.
Precipitation (with nitric acid and silver nitrate)	No white precipitate	Sample doesn't contain chloride ions.
Carbonate Test	Limewater turned cloudy	Sample contains a carbonate.

CONCLUSION: The white powder is likely to contain <u>sodium carbonate</u>, which is commonly found in washing powder.

Signature: Date: 14 / 09 / 2006

To make the results more <u>reliable</u>, they should be repeated at least once, by a different scientist. The sample should also be kept free from <u>contamination</u> (see p.36).

Instrumental Techniques are Better than Simple Experiments

In reality, none of this actually happens any more. There are <u>faster</u>, more <u>accurate</u> and more <u>reliable</u> techniques for working out what substance you have — things like <u>mass spectrometry</u> and <u>infrared analysis</u>. These techniques can work with <u>tiny quantities</u> and give more <u>detailed readings</u>.

But you have to know all the <u>simple laboratory experiments</u> anyway — life's just like that sometimes.

Sherlock Holmes never looked so good in a lab coat...

So if you ever come across a suspicious white powder, you'll know what to do...

Revision Summary for Section 2.5

My favourite part of any section — the end. But it's not over yet for all you budding crime investigators — there's still this beautiful revision summary to get through. You've covered ionic and covalent compounds and a handful of tests which you need to be able to describe. If your memory fails you, then have a look back through the section. If you forget a single one you might miss a vital clue in the investigation — and how else are you going to work out who did it?

1) Give two things that determine the properties of a substance.

2) What type of structure do ionic compounds have?

3) Explain why ionic compounds are usually solids at room temperature.

4) Are the following ionic compounds soluble in water? a) potassium nitrate, b) lead sulfate, c) ammonium chloride, d) sodium hydroxide, e) calcium carbonate.

5)* Use the table to help you find the formula for:
    a) iron(II) oxide, b) iron(III) oxide,
    c) calcium chloride, d) sodium carbonate.

6) Are substances found in living materials mostly ionic or covalent?

Positive ions		Negative ions	
sodium	Na^+	chloride	Cl^-
calcium	Ca^{2+}	oxide	O^{2-}
iron(II)	Fe^{2+}	carbonate	CO_3^{2-}
iron(III)	Fe^{3+}		

7) Explain why covalent substances are usually gases or liquids at room temperature.

8) Give the formulas of the following compounds: a) carbon dioxide, b) ethanol, c) glucose.

9) Describe how to carry out a flame test.

10) What ions are present if a flame test produces a: a) lilac flame? b) blue-green flame?

11) What is universal indicator?

12) A substance was found to have a pH of 5. Is it an acid or an alkali?

13) Describe how you could test to see if a compound is soluble in water.

14) Describe how you could make a solution from a solid sample (that's soluble in water).

15) What gas is released when a carbonate is placed into some dilute acid? How could you test for this gas?

16) Describe how you could distinguish between solutions of: a) calcium sulfate and copper sulfate, b) copper nitrate and copper sulfate, c) sodium chloride and sodium hydroxide.

17)* Name the product that's formed during each of the following reactions:
     a) barium chloride + iron(II) sulfate, b) silver nitrate + zinc chloride

18) What colour change occurs when ethanol is passed over crystals of acidified potassium dichromate?

19) Describe the original breathalyser, and suggest why it isn't very accurate.

20) Describe how you could find out if a substance contains glucose.

21)* Forensic scientists are investigating the identity of an unknown white solid. They conduct a series of chemical tests and obtain the following results:
     • soluble in water,
     • produces a blue precipitate with sodium hydroxide solution,
     • produces a white precipitate with barium chloride solution (after first adding dilute HCl).
     a) Name the white solid.
     b) Give the formula of this compound.

22) Give three advantages of instrumental techniques compared to simple laboratory experiments.

* Answers on p.100.

Comparison Microscopes

Some pieces of evidence are <u>way too small</u> to look at and analyse with the naked eye, which is where microscopes come in... but don't worry — it's not all technical gubbins, there's some cool stuff too...

A Comparison Microscope is Used to Compare Things

1) A comparison microscope is a bit like <u>two light microscopes</u> (like you'd use at school) <u>stuck together</u>.

2) One bit of evidence is put <u>under one side</u> and another bit of evidence (the one you're comparing it to) is put <u>under the other side</u>.

3) When you look in the eyepiece you see the two bits of evidence right <u>next to each other</u>. This photo shows a forensic scientist using one to compare bullets.

4) This makes it loads <u>easier</u> to compare the two to see if you have a <u>match</u> (you don't have to keep swapping microscope slides).

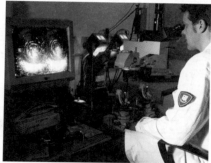

MAURO FERMARIELLO / SCIENCE PHOTO LIBRARY

They're Used to Compare Bullets and Seeds

They're good for looking at <u>bullets</u> and <u>seeds</u> (see below), and also for looking at <u>fibres</u> and <u>soil</u> (see next page). You need to know the <u>distinctive features</u> that allow forensic scientists to <u>match</u> these pieces of evidence.

BULLETS

1) Bullets are sometimes found at the scenes of <u>violent crimes</u>.

2) They're useful in forensics because fired bullets are <u>unique</u> to <u>one gun</u>.

3) They can be used to <u>match guns to crimes</u>, and if you can then match a person to a gun you've found a possible suspect. Genius.

1) <u>Length</u> of bullet.

2) The <u>shape</u> and <u>weight</u> of the bullet can tell you the <u>calibre</u> of the cartridge (which helps identify the <u>type</u> of gun).

3) <u>Rifling marks</u> on bullet. Gun barrels have tiny imperfections inside. When a bullet passes along the barrel, it gets scraped and lines are produced on the bullet. These marks are <u>unique</u> — no two guns produce the same marks.

4) <u>Other objects</u>. Bits of <u>glass</u> can be embedded in bullets (e.g. from the bullet going through a window). If the bullet goes through <u>cloth</u> (e.g. if someone is shot through their clothing) the cloth can <u>leave a pattern</u> on the bullet.

SEEDS

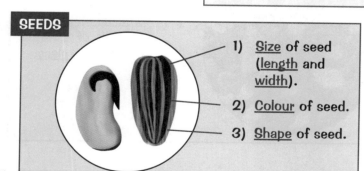

1) <u>Size</u> of seed (<u>length</u> and <u>width</u>).

2) <u>Colour</u> of seed.

3) <u>Shape</u> of seed.

1) Seeds might seem a <u>bit dull</u> but they can be used to show <u>where</u> objects or people <u>have been</u> (just like pollen can — see p.53).

2) This information can be used to <u>back up alibis</u>, or <u>link</u> people or objects to <u>crime scenes</u>, e.g. if the same seeds that are present at the crime scene are found on a suspect's trousers then there may be a link.

Comparison microscopes — mine's bigger than yours...

Don't forget you're comparing <u>evidence from the crime scene</u> to <u>something from the suspect</u>.

Polarising Microscopes

Sometimes a light microscope <u>doesn't show up everything</u> though. Forensic scientists can use a nifty bit of equipment called a polarising microscope to reveal more...

Polarising Microscopes are a Bit Different

1) Polarising microscopes are <u>kinda complicated</u>, but luckily you don't need to know how they work.

2) They're used because they <u>let you see things</u> that you <u>can't see</u> using a <u>light microscope</u>.

3) Forensic scientists will still use a comparison microscope too, but polarising microscopes can help reveal extra details in the evidence.

They're Good For Looking at Fibres...

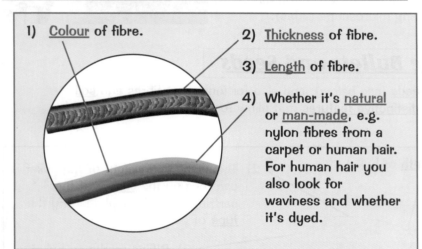

1) <u>Colour</u> of fibre.

2) <u>Thickness</u> of fibre.

3) <u>Length</u> of fibre.

4) Whether it's <u>natural</u> or <u>man-made</u>, e.g. nylon fibres from a carpet or human hair. For human hair you also look for waviness and whether it's dyed.

1) Fibres are <u>found</u> at all sorts of <u>different crime scenes</u>, e.g. a few fibres from a burglar's jacket on a broken window, a human hair from an attacker on a victim's shirt etc.

2) Fibres <u>aren't usually unique</u>, but they can still be used to <u>link</u> a suspect (or their clothing, or belongings) to a crime scene.

3) The sort of fibres forensic scientists might look at are human hairs, animal hairs, clothing fibres and carpet fibres.

4) Using a <u>polarising</u> microscope lets you see <u>man-made fibres</u> much more clearly.

...and Soil

1) Soil is made up of <u>minerals</u>, bits of <u>rock</u>, <u>organic matter</u> (e.g. dead leaves and worms) and <u>water</u>.

2) Soil <u>composition varies</u> a lot from place to place. This makes it pretty useful evidence in forensics because it can show <u>where</u> something has been, e.g. soil on a suspect's shoe could be linked to a muddy crime scene.

3) Using a <u>polarising</u> microscope makes it easier to see things like <u>glass fragments</u> and <u>minerals</u> in the soil sample.

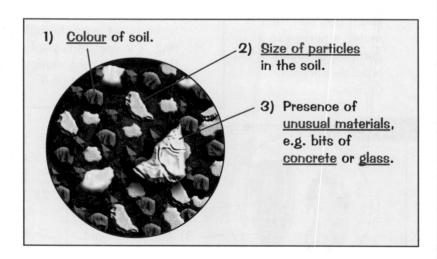

1) <u>Colour</u> of soil.

2) <u>Size of particles</u> in the soil.

3) Presence of <u>unusual materials</u>, e.g. bits of <u>concrete</u> or <u>glass</u>.

How do you polarise a microscope? Take it to the Arctic...

Sorry... lame joke. Anyway, fibres and soil, hmmm... they're not the most thrilling things in the world but they can provide <u>strong evidence</u> of a link between a suspect and a crime scene. Just think — your muddy shoes could prove you've been down the park when you were supposed to be in bed. _{Unlucky for you if your mum's a forensic scientist.}

Electron Microscopes

Some distinctive features are too small to even see on a light microscope (they're really, really, really small) — when this is the case scientists use electron microscopes...

Electron Microscopes are Very Powerful

1) Electron microscopes use a <u>beam of electrons</u> (tiny charged particles) to <u>produce an image</u> of the evidence on a computer screen — called an electron micrograph. (You can't actually look down an eyepiece and see what's under the microscope.)

2) They can <u>magnify</u> images many more times than light microscopes, so objects appear <u>larger</u>.

3) They have a <u>higher resolution</u> than light microscopes, so you can see <u>more detail</u>.

4) One of the drawbacks is that it <u>can't show colour</u> (if you see a coloured electron micrograph it's because someone's used a computer to colour it in, just to make it a bit clearer). This means evidence is usually looked at with <u>both</u> a light microscope and an electron microscope.

5) Electron microscopes are pretty <u>expensive</u> though so you won't find one at school.

They're Good for Looking at Layers of Paint...

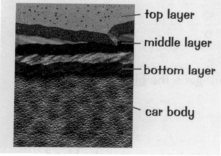

This is what an electron micrograph of three layers of paint on a bit of car body might look like.

- top layer
- middle layer
- bottom layer
- car body

1) The paint forensic scientists are most likely to look at is the paint on <u>cars</u>. When a car's painted <u>lots</u> of <u>thin layers</u> are applied.

2) If scientists know the <u>colour</u> and <u>number</u> of <u>different layers</u> then they can narrow down the <u>make</u>, <u>model</u> and <u>age</u> of the car the paint came from.

3) This is really useful for investigating <u>car crashes</u> and <u>hit-and-runs</u> — flecks of paint are transferred onto other cars or a victim's clothing when they hit them and can be used to track down the suspect.

4) Electron microscopes can show the different layers of paint <u>much better</u> than light microscopes because the electrons reflect differently off <u>different types of paint</u>.

...And Pollen Grains

1) Pollen grains can tell you a lot about <u>where</u> an object or person has <u>been</u> because <u>different pollens</u> are found in <u>different areas</u>. This can help to do loads of things like <u>link</u> suspects to a crime scene or figure out which country things like counterfeit money have come from.

For example, a woman was arrested for suspected arson. Forensic scientists found a mixture of pollens on her trouser legs. This matched the mixture of pollens in the back garden of the house that burnt down — linking the woman to the crime scene.

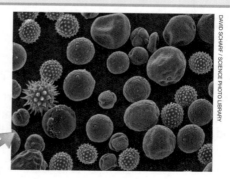

DAVID SCHARF / SCIENCE PHOTO LIBRARY

2) When forensic scientists are comparing pollens they look at the <u>size</u>, <u>shape</u> and <u>surface structure</u> of pollen grains. Each <u>type</u> of plant has <u>distinctive pollen</u>.

3) Electron microscopes are used to look at pollen grains because they're <u>really small</u> (around a tenth of a millimetre). Also, electron microscopes can give a more <u>detailed image</u> of the <u>pollen surface</u>.

I think I prefer layers of cake, to be honest...

Paint layers on cars usually <u>aren't unique</u> (but they can be if the car has been custom painted). Pollen grains are unique to each plant, so the mixture of pollens can be really useful.

Blood Group Typing

Blood is often found at the scenes of <u>violent</u> crimes, e.g. murders and assaults, hmmm... gruesome.
<u>Blood typing</u> can tell forensic scientists whether a blood sample is <u>human</u> and which <u>blood group</u> it is.
The blood group can then be <u>compared</u> to the <u>suspect's</u> to see if they <u>match</u>.

Blood Contains Four Main Things

1) Red Blood Cells	2) White Blood Cells	3) Platelets	4) Plasma
They <u>carry oxygen</u> to all the cells in the body. They make the blood look <u>red</u>.	Their main role is <u>defence against disease</u>.	These are small fragments of cells that help the <u>blood to clot</u>.	This is the <u>liquid</u> that <u>carries</u> everything about.

There are Four Main Blood Groups

1) People have different <u>blood groups</u> (sometimes called blood <u>types</u>)
   — you can be one of: <u>A</u>, <u>B</u>, <u>O</u> or <u>AB</u>.

2) These letters refer to the type of <u>antigens</u> on the surface of a person's <u>red blood cells</u>. (An antigen is a substance that can trigger a response from a person's <u>immune system</u>.)

3) Red blood cells can have <u>A or B antigens</u> (or <u>neither</u>, or <u>both</u>) on their surface.

4) And blood plasma can contain <u>anti-A or anti-B antibodies</u> (antibodies are chemicals produced by the immune system).

Blood Group	Antigens	Antibodies
A	A	anti-B
B	B	anti-A
AB	A, B	none
O	none	anti-A, anti-B

5) If <u>anti-A</u> antibodies <u>meet A</u> antigens OR <u>anti-B</u> antibodies <u>meet B</u> antigens the <u>blood will clot</u>.

You Can Test for Blood Group

When a <u>red stain</u> or substance is found at a crime scene scientists need to <u>test</u> whether it's <u>blood</u> (not, e.g. dye or ketchup). They then need to <u>test</u> if it's <u>human blood</u> (not blood from the dog or cat etc.) <u>Chemicals</u> are used to test for blood, then <u>anti-human antibodies</u> show if the blood is <u>human</u>.
Once they've figured out it's human blood they can then test for the <u>blood group</u>:

1) Scientists can test for blood group by <u>mixing different antibodies</u> with <u>blood samples</u>.

2) Depending on whether the blood <u>clots</u> or <u>not</u> they can tell which blood type it is.

Blood Group	Antigens	Does it clot with anti-A antibodies?	Does it clot with anti-B antibodies?
A	A	Yes	No
B	B	No	Yes
AB	AB	Yes	Yes
O	none	No	No

E.g. if scientists add some <u>anti-A antibodies</u> to a blood sample and <u>it clots</u> then the blood group must be <u>either A</u> or <u>AB</u> (because it would only clot if <u>A</u> antigens were <u>present</u>). If they then added <u>anti-B antibodies</u> to the same blood and it didn't clot they would know it was blood <u>group A</u> (because if it had been AB then it would clot with anti-B antibodies).

If the blood group of the blood sample from the crime scene and the suspect's blood group <u>match</u> then your suspect <u>is still a suspect</u>. You <u>can't</u> say it's <u>definitely their blood</u> because there are only four blood types — so there are 1000s of other people that have the same blood type.

Blood typing — gets your keyboard a bit messy...

Nowadays blood typing <u>isn't used</u> in forensics because of <u>DNA profiling</u> (see next page) — blood typing only <u>narrows down</u> the list of possible suspects, whereas DNA profiling can tell you if the blood came from a particular person. Blood typing is still used when doing blood transfusions though.

DNA Profiling

Everywhere you go you <u>leave</u> a <u>trail</u> of your <u>DNA behind</u> — e.g. in skin flakes on clothing, hair in your hairbrush etc. And unless they're really careful, <u>criminals</u> usually leave DNA at the <u>crime scene</u>, which is pretty handy for forensic scientists — they use DNA profiling to figure out who it belongs to...

DNA is Unique

1) Remember that DNA is the <u>genetic material</u> found in the <u>nucleus</u> of your <u>cells</u>. It's a bit like a <u>blueprint</u> for how to make a human being.

2) Your DNA is <u>unique</u> — no one else in the world has the same DNA as you (unless you're an <u>identical twin</u>, then the two of you have <u>identical DNA</u>).

3) DNA can be <u>extracted</u> from hair, skin flakes, <u>blood</u>, <u>semen</u> and <u>saliva</u> because they all <u>contain cells</u>.

4) <u>DNA profiling</u> (or genetic fingerprinting) is a way of <u>comparing DNA samples</u> to see if they come from the same person or from two different people.

DNA Profiling Pinpoints Individuals

1) DNA taken from a <u>crime scene</u> is usually compared with a DNA sample taken from a <u>suspect</u>. If there isn't a suspect, the police enter it into a <u>national database</u> (see p.59) to search for a <u>match</u>.

2) It can also be used in <u>paternity tests</u> — to check if a man is the father of a particular child. This is because children <u>inherit</u> some of their DNA from their <u>mum</u> and some from their <u>dad</u> — so their DNA profiles will be <u>similar</u> to those of their parents.

HOW IT WORKS

1) First you have to <u>extract</u> the DNA from the <u>cells</u> in the blood, semen etc.

2) The <u>DNA</u> is then <u>cut up</u> into <u>fragments</u>.

3) This produces lots of <u>different sized bits</u> of DNA. The number of each size will be <u>different for everyone</u> because of the way it's cut.

4) The DNA bits are <u>separated</u> using a process called <u>electrophoresis</u> (a bit like chromatography — see p.58). They're <u>suspended in a gel</u>, and an <u>electric current</u> is passed through the gel. DNA is <u>negatively charged</u>, so it moves towards the <u>positively charged end</u> of the gel (because charged particles move in an electric field). Small bits travel <u>faster</u> than big bits, so they get <u>further</u> through the gel.

5) The DNA is then <u>treated</u> to make it <u>visible</u>.

DNA moves towards the positive end, with smallest fragments moving furthest

Here's an example:

1) A drop of <u>blood</u> was found at a <u>crime scene</u>.

2) Forensic scientists ran a <u>DNA profile</u> for the blood.

3) They also ran DNA profiles for <u>two suspects</u>.

4) <u>Matching</u> DNA samples have the <u>same pattern</u> of bands. So here you can see that the blood from the crime scene has come from <u>suspect 2</u>.

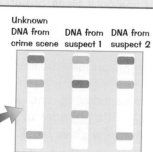

So the trick is — frame your twin and they'll never get you...

In the exam you might have to interpret data on <u>DNA fingerprinting for identification</u>. They'd probably give you a diagram similar to the one at the bottom of this page, and you'd have to say <u>which</u> of the <u>known</u> samples (if any) <u>matched</u> the <u>unknown</u> sample. Pretty easy — it's the two that look the same.

Identifying Glass: Blocks

Identifying broken bits of <u>glass</u> and <u>plastic</u> can be really handy for tracking down suspects...

Glass and Plastic Can Refract Light

1) When <u>light</u> goes from one <u>substance</u> (or <u>medium</u>) into <u>another substance</u>, e.g. from air into glass, it gets <u>refracted</u>.

2) Refraction is when waves <u>change direction</u> as they enter a different medium.

3) This is caused entirely by the change in <u>speed</u> of the light waves.

You can't fail to remember the old '<u>ray of light through a rectangular glass block</u>' trick:

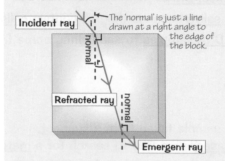

The 'normal' is just a line drawn at a right angle to the edge of the block.

1) The '<u>incident</u>' ray is just the ray hitting the block.

2) It bends <u>towards the normal</u> as it enters the <u>denser medium</u> (the glass), and <u>away</u> from the normal as it <u>emerges</u> into the <u>less dense</u> medium (the air).

3) The angle between the normal and the incident ray is called the <u>angle of incidence</u>, <u>i</u>.

4) The angle between the normal and the refracted ray is called the <u>angle of refraction</u>, <u>r</u>.

Glass Can be Identified Using Its 'Refractive Index'

1) There are lots of <u>different types</u> of glass and clear plastic, and they all <u>bend</u> light by <u>different amounts</u>.

2) <u>How much</u> a material bends light is called its '<u>refractive index</u>'.

3) The <u>higher</u> the refractive index, the <u>more</u> the light is bent as it passes from air into the material.

4) You can work out the refractive index, <u>n</u>, of a material using <u>Snell's Law</u>.

Snell's Law Says...

<u>When a light ray passes into a material:</u>

So if you know <u>i</u> and <u>r</u>, you can work out the <u>refractive index</u>.

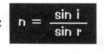

$$n = \frac{\sin i}{\sin r}$$

(Thankfully you don't have to know <u>why</u> Snell's law works. Just that it does.)

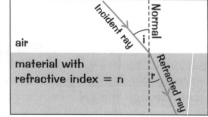

TO FIND THE ANGLES OF INCIDENCE AND REFRACTION:

1) Draw a <u>straight line</u> on a piece of paper — this will be your '<u>normal</u>'.

2) Place the block of glass or plastic carefully at <u>right angles</u> to the normal.

3) Shine a <u>fine beam</u> of light at the block so that it meets the block at an <u>angle</u> to the normal.

4) Using a <u>protractor</u>, carefully measure the angles of <u>incidence</u> and <u>refraction</u>.

5) So that's what 'refractive index' <u>is</u>, and you can use this method to work out the refractive index of a glass or plastic <u>block</u>. It's <u>not</u> a lot of good for <u>criminal evidence</u> though (unless someone's been nicking glass blocks from the glass block factory) — you tend to get <u>small shards</u> of glass instead. You need to use a <u>different method</u>, which is covered on the next page.

Revise this — and make light work of it in the exam...

The <u>basic ingredients</u> of all glass are sodium carbonate, sand and limestone, but most common types of glass have <u>added ingredients</u> that affect the refractive index. For example, lead is added to some glass to make 'lead crystal'. This <u>increases</u> the refractive index of the glass and makes it 'sparklier'.

Identifying Glass: Fragments

Once you know the refractive index of the glass from a <u>crime scene</u>, you can compare it with shards of glass found on the <u>clothes</u> or <u>shoes</u> of a suspect.

Most of the bits of glass you find on a suspect are <u>less than a millimetre</u> across though, which is a <u>wee bit small</u> for the method on the previous page — you'd have to have a <u>mighty small protractor</u>...

Glass Fragments <u>are</u> <u>Too Small</u> to Use Snell's Law

1) You can only use <u>Snell's law</u> (previous page) to find the refractive index of glass if the piece of glass is <u>big enough</u> — you need to be able to see the refracted beam clearly so you can measure the angle of refraction.

2) For <u>small pieces</u> or <u>shards</u> of glass you need to use this clever little fact:

> If you have <u>two materials</u> next to each other with the <u>same</u> refractive index, you <u>can't see the boundary</u> between them. So, a piece of glass in a liquid of the same refractive index would be <u>COMPLETELY INVISIBLE</u>. (Oooooh...)

You Use the <u>Oil Immersion Temperature</u> Method Instead

Some <u>oils</u>, e.g. <u>silicone oil</u>, have a refractive index that <u>changes</u> with <u>temperature</u>. You can vary the temperature of the oil until it has the same refractive index as your glass sample. Here's how it works:

1) Using <u>tweezers</u>, carefully place your tiny glass sample onto a <u>microscope slide</u>.

2) <u>Cover</u> the sample with a <u>few drops</u> of silicone oil (a clear, colourless liquid) and close the slide with a cover slip.

3) Push your slide into a piece of apparatus called a <u>hotstage</u>.

4) Then put the whole thing under a <u>light microscope</u> (making sure the glass and the oil around it are well lit) and focus on the <u>boundary</u> between the glass and the oil.

<u>HOTSTAGE</u>: used to warm the oil on the slide slowly and evenly.

electric heater

hole to let light through from below

slot for slide

heater elements above and below slide

5) <u>Slowly heat</u> the oil using the hotstage.

6) As the <u>temperature</u> of the oil increases, it's refractive index <u>drops</u>. At a certain temperature, the boundary will <u>disappear</u> — the oil has the same refractive index as the glass.

You can then <u>look up</u> the refractive index of silicone oil at that temperature to find the refractive index of your glass sample.

In <u>modern</u> forensic labs, the observing is done by a <u>camera</u> attached to a <u>computer</u>. This is more <u>accurate</u> than observing the boundary by eye, since it doesn't rely on <u>human judgement</u>.

Hotstage — that's got to sting on panto night...

Of course, even if your bit of glass from the crime scene <u>matches</u> glass on the suspect, that <u>doesn't</u> mean they committed the crime. If the glass is <u>common</u>, the suspect could have picked it up somewhere else entirely. Or they could have been at the crime scene <u>without</u> committing the crime. Tricky...

Chromatography

Chromatography is used a lot in the chemical industry — it's a <u>method</u> for <u>separating chemical mixtures</u>. It's pretty useful in <u>forensics</u> because you can use it to do cool things like <u>compare ink samples</u> to <u>detect forgeries</u> (fake documents).

Chromatography Can be Used to Detect Forgeries

Most <u>inks</u> are made up of a <u>mixture of dyes</u>. A forged document will probably use <u>different ink</u> from an <u>official document</u> (so it'll contain a different mixture of dyes). <u>Different dyes</u> in ink will <u>wash</u> through paper at <u>different rates</u>. Dyes move up the paper at <u>different rates</u> because they have <u>different solubilities</u>. The <u>more soluble</u> chemicals move up the paper <u>faster</u> than the less soluble ones. Chromatography uses this <u>property</u> to separate out the component dyes.

Using Paper Chromatography...

Here's how you do it...

1) Draw a <u>line</u> across the bottom of a sheet of <u>filter paper</u> (in pencil).

2) Add <u>spots of ink</u> to the line at regular intervals.

3) Tape the top of the paper to a pencil and <u>hang</u> the sheet in a <u>beaker of solvent</u>, e.g. <u>water</u>.

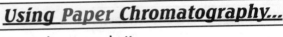
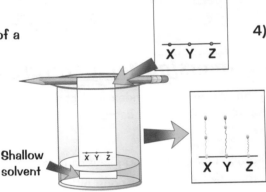

Shallow solvent

4) The solvent <u>seeps</u> up the paper, carrying the ink dyes with it.

5) Each different dye will move up the paper at a <u>different rate</u> and form a <u>spot</u> in a different place.

6) You can <u>compare</u> the dyes in an unknown ink to the dyes in <u>known inks</u> to see which ink it is. The <u>pattern</u> of dye spots will match when two inks are the same (the spots will be the same distance apart).

EXAMPLE:

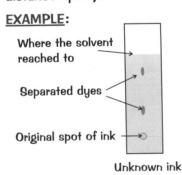

Where the solvent reached to

Separated dyes

Original spot of ink

Unknown ink

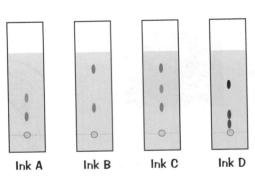

Ink A   Ink B   Ink C   Ink D

You can see from the <u>position</u> of the <u>spots</u> on the filter paper that the unknown ink has the <u>same composition</u> as <u>ink B</u>.

...or Thin-Layer Chromatography

Thin-Layer Chromatography (TLC) is very similar to paper chromatography. The main difference is that instead of paper it uses a <u>thin layer of gel or paste</u> (e.g. silica gel) on a <u>glass plate</u>. It's <u>faster</u> than paper chromatography and gives <u>better separation</u>.

A bit of TLC always cheers me up...

Chromatography isn't just great for detecting dodgy tenners, e.g. it can be used to help track down weirdos who send <u>threatening letters</u>. The ink in a printed letter could be analysed and compared to the ink from suspects' printers — this'll tell you which printer the letter could have been made on (but not who made it).

Using Databases and Records

OK, this page sounds like it might be about as exciting as watching paint dry, but it's not all bad (and you have to learn it anyway), so bear with me...

The Police Can Use Databases and Records to Find Information

There are several national databases and sources of records that police and forensic scientists use to find information to help with their investigations. Here are some of the things they use:

1) **FINGERPRINT RECORDS**   There's a <u>central database</u> of <u>fingerprints</u> that police forces can search. Only people who have had their prints taken during a criminal investigation have their fingerprints stored though — <u>not everyone</u> in the country is on it. The database is used to do things like <u>find a suspect</u> from fingerprints left at a crime scene and help <u>identify dead bodies</u>.

2) **VEHICLE RECORDS**   The Driver and Vehicle Licensing Agency (DVLA) have a <u>database</u> of <u>driving licenses</u> and <u>vehicle details</u>. This database is used to search for cars (unsurprisingly) or owners of cars, e.g. when looking for the owner of a getaway vehicle.

3) **DENTAL AND MEDICAL RECORDS**   Dental records are often used to <u>identify skeletons</u> and <u>disfigured bodies</u>. Medical records might be used by the police if they know their suspect suffers from, say, a genetic disorder or diabetes.

4) **DNA RECORDS**   There's a <u>National DNA Database</u> (NDNAD) that police forces can search. It holds DNA records from unsolved crimes, from suspects and from convicted criminals. This is normally searched to <u>match DNA from a crime scene</u> to a suspect.

5) **MISSING PERSON RECORDS**   The police have records of missing people. These are usually searched when trying to <u>identify a dead body</u>.

6) **VALUABLE ITEM RECORDS**   <u>Insurance companies</u> hold records of valuable items. These are searched to find out things like where stolen items have come from or if an item has been recorded as stolen.

If your suspect <u>is in a database</u> and you get <u>no match</u> when you search it you can <u>sometimes exclude</u> that suspect from your investigation. It's very hard to <u>completely</u> rule someone out though — just because it isn't the suspect's DNA, fingerprints etc. doesn't mean they weren't involved. There are loads of ways not to leave evidence, e.g. by wearing gloves to avoid leaving fingerprints.

Witness Descriptions Also Help

1) In a lot of crimes <u>there are witnesses</u> (people who see it happening or see the suspect). Witness descriptions are taken and turned into a <u>drawing</u> or a <u>picture</u> of the <u>suspect</u>. This helps police when they're looking for the suspect, e.g. they may compare it to known criminals or show it on the news.

2) There are <u>forensic artists</u> that work with the witness to create a drawing of the suspect.

3) Sometimes the police use an <u>Identikit</u>. Witnesses are shown lots of <u>versions of a feature</u>, e.g. eyes, and <u>pick</u> the most similar to the suspect. They repeat this for all the facial features, e.g. nose, ears, lips etc. to give a picture of the suspect's face.

MAURO FERMARIELLO / SCIENCE PHOTO LIBRARY

Identikit — Mr Potato Head for suspects...

A police investigation might use loads of different databases to find suspects or gather more evidence against them. Forensic scientists tend to use fingerprint and DNA databases more than any others.

Links to a Crime Scene

Drawing <u>conclusions</u> from evidence is a <u>tricky business</u> — you have to be careful not to go accusing someone of murder if all the evidence shows is that they were at the crime scene (they won't be happy — trust me).

A Link Doesn't Mean They Did It

1) If the hair, blood, bullet, footprints, fingerprints etc. found at the crime scene <u>match</u> that from the suspect or their property, it <u>doesn't mean</u> they <u>committed the crime</u>.

2) There could be <u>some other explanation</u> for the evidence being there, e.g.

- The suspect had <u>been</u> to the <u>crime scene</u> on <u>another occasion</u>.
- Someone <u>planted</u> the evidence to incriminate the suspect.
- Something used in the crime was <u>stolen</u> from the suspect.

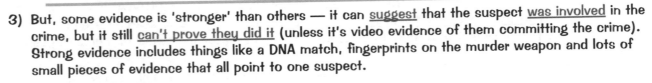

3) But, some evidence is 'stronger' than others — it can <u>suggest</u> that the suspect <u>was involved</u> in the crime, but it still <u>can't prove they did it</u> (unless it's video evidence of them committing the crime). Strong evidence includes things like a DNA match, fingerprints on the murder weapon and lots of small pieces of evidence that all point to one suspect.

Example 1 — a Fairly Clear Case

A man is found beaten and <u>shot dead</u> in a field. Forensic scientists <u>collect</u> the following things:

From the <u>crime scene</u>:
- A <u>bullet</u> (the one that killed the man).
- <u>Red man-made fibres</u> from the dead man's jacket.

From a <u>suspect's car</u>:
- A <u>gun</u> with <u>bloody fingerprints</u> on it.
- A <u>rug</u> made of <u>red man-made fibres</u>.

Then they <u>analyse</u> them:
The fibres on the dead man's jacket did match those from the rug in the suspect's car (so the man might have been in the suspect's car). The rifling marks on the bullet matched the rifling marks on a bullet test-fired from the gun found in the suspect's car. The fingerprints on the gun also match the suspect's and the blood they are made in is from the victim.

In this case there's a <u>high probability</u> the suspect <u>is linked</u> to the crime scene. But, it doesn't prove they killed the man — there are still other explanations for the evidence, e.g. the suspect found the man and moved the gun whilst trying to help him.

Example 2 — Hmm... Not so Sure

A girl is a victim of a <u>hit-and-run</u>. Forensic scientists <u>collect</u> the following things:

From the <u>crime scene</u>:
- Flecks of <u>paint</u> from the girl's clothing.

From a <u>suspect's car</u>:
- A <u>paint sample</u>.

Then they <u>analyse</u> them:
The flecks of paint on the girl's clothing did match the paint sample taken from the suspect's car.

In this case there's a <u>lower probability</u> the suspect <u>is linked</u> to the crime scene. Many cars have the exact same paint layers so thousands of people could have done it. <u>Without more evidence</u> you can't be confident that the suspect did it.

It was Professor Plum in the kitchen with the candlestick...

It's the <u>police</u>'s job to find the suspects, <u>based on information</u> from <u>witnesses</u> and the <u>forensic scientists</u>. The forensic scientists also rule out suspects. A suspect is arrested if there's enough evidence and then the courts decide (based on the evidence) if they're guilty. It's like a giant whodunnit...

Forensic Science

Revision Summary for Section 2.6

Well, I bet after all that you fancy yourself a bit of a Miss Marple (how can I blame you — she has such marvellous fashion sense). But, to be a forensic scientist you need to know your stuff and not just accuse someone because they have a sinister beard, shifty walk or a white cat...

1) What is a comparison microscope? How is it used in forensic science?
2) Give three distinctive features of a fired bullet.
3) Give three distinctive features of a seed.
4) For what things is a polarising microscope sometimes more useful than an ordinary light microscope?
5) Give four features that forensic scientists look at when comparing fibres.
6) Give three features that forensic scientists look at when comparing soil. Why might you look at soil with a polarising microscope?
7) Give two advantages and two disadvantages of electron microscopes.
8) What do forensic scientists look for when comparing layers of paint?
9) Why do forensic scientists use pollen grains as evidence? What do they look for when comparing pollen samples?
10) What are the four main components of blood?
11) What are the four blood groups?
12) What kind of red blood cell antigens will a person with blood type A have?
13) What kind of antibodies will a person with blood type A have in their plasma?
14) Describe how scientists test for blood group.
15) A scientist tests a blood sample for blood group. They add anti-A antibodies and the blood does not clot. They then add anti-B antibodies and the blood does clot. What blood type is the sample?
16) Where in a cell is DNA found?
17) Describe the method used to make a DNA profile.
18) Draw a diagram to show the path of a ray of light as it passes from air → block of glass → air, meeting the block of glass at an angle to the normal. Mark on your diagram the angles of incidence and refraction.
19) Describe how you would find the refractive index of a glass block.
20) Explain why you can't usually use the method you described in the answer to question 19 to find the refractive index of glass from a crime scene or suspect.
21) What method can forensic scientists use to find the refractive index of small pieces of glass? Explain how this method works.
22) Give one use of chromatography in forensic science.
23) Name two different types of chromatography.
24) Why might the police search fingerprint records?
25) Why might the police search DNA records?
26) Can you exclude suspects from your investigation based on the results of a search?
27) Give two methods of recording a witness description.
28) A man's gun was found at a crime scene. The police think he might have committed the crime. Give two other explanations for how the gun got there.

Success in Sport

If you want to be at the <u>top</u> of your sport, it's not just about the <u>obvious</u> things like fitness and talent. You also need to look at your <u>diet</u>, <u>equipment</u>, <u>clothing</u> and <u>mental attitude</u> as well.

Success in Sport Depends on Lots of Things

1) How Fit You Are

Being successful in sport depends a lot on how <u>physically fit</u> you are — your body needs to be able to <u>perform</u> well when you're putting it under <u>stress</u> in competitive situations.

For most sports, you need <u>cardiovascular fitness</u>, <u>muscular strength</u>, or <u>both</u>:

1) <u>CARDIOVASCULAR FITNESS</u> — your <u>heart</u> needs to be able to pump oxygen around your body quickly, so that your muscle cells get enough oxygen for respiration (see p.66). You also need to be able to <u>recover</u> quickly after exercise.

2) <u>MUSCULAR STRENGTH</u> — you need to develop the <u>muscles</u> that are needed for your sport, e.g. strong arms for tennis players.

This section is all about <u>understanding</u>, <u>measuring</u> and <u>improving</u> physical fitness for sport.

2) The Food You Eat

Your <u>energy</u> and <u>nutrient</u> intake can make a real difference to how you perform in sport.

1) Athletes often eat <u>specific things</u> to give them the energy they need for their sport, or to help build muscle mass — e.g. body-builders might eat a high protein diet.

2) Advising athletes what to eat is the role of the <u>sports nutritionist</u> (see Section 2.8).

3) The Equipment You Use

In most sports, the <u>equipment</u> and <u>clothing</u> that you use can make a big difference to your performance.

1) E.g. the aerodynamic <u>helmets</u> some racing cyclists use. Being super-fit isn't enough — you have to reduce <u>air resistance</u> as well if you want to scrape extra seconds off your time.

2) Finding the right materials for the job is the role of a <u>materials scientist</u> (see Section 2.9).

4) How Skilled You Are

Bit of an obvious one this, but if you don't have <u>good technique</u> you'll never play very well.

1) E.g. to be a footballer, you need good <u>ball control</u> as well as speed and agility.

2) To succeed in just about any sport, you need a wee bit of <u>natural talent</u> and an awful lot of <u>hard work</u> to learn and practise the best techniques.

5) How You Perform Under Pressure

None of that will do you any good if you can't <u>concentrate</u> and <u>focus</u> in a competitive situation.

1) This can make the difference between winning and losing — sometimes the <u>most skilled</u> athlete doesn't win because they haven't got the right <u>mental attitude</u>.

2) A <u>good trainer</u> can help you work on your attitude and performance in competitive situations.

My trainer gets my foot wet...

It's not just a simple matter of being fit for a professional athlete — you've got to eat the right stuff, use the right equipment, be skilled AND perform under pressure. But I guess you get to be famous...

The Cardiovascular System

It's the job of a sports physiologist to help athletes improve their physical fitness and strength. Before they can do that, they have to understand how the various bits of the body work together. And guess what... so do you. Read on.

Blood is Carried Around the Body by Vessels

There are three different types of blood vessel:

1) **ARTERIES** — these carry the blood away from the heart.

2) **CAPILLARIES** — these are tiny vessels where materials are exchanged at the tissues.

3) **VEINS** — these carry the blood to the heart.

Normally, arteries carry oxygenated blood (blood with oxygen) and veins carry deoxygenated blood (blood without oxygen).

The pulmonary artery and pulmonary vein are the big exceptions to this rule (see diagram below).

The Blood is Pumped Around the Body by the Heart

1) The heart is a pump. It supplies the force needed to push the blood all round the body through the blood vessels. To every last tissue and back.

2) Humans have a double circulatory system. This means that there are two circuits of blood vessels — one going to the lungs and one to the rest of the body:

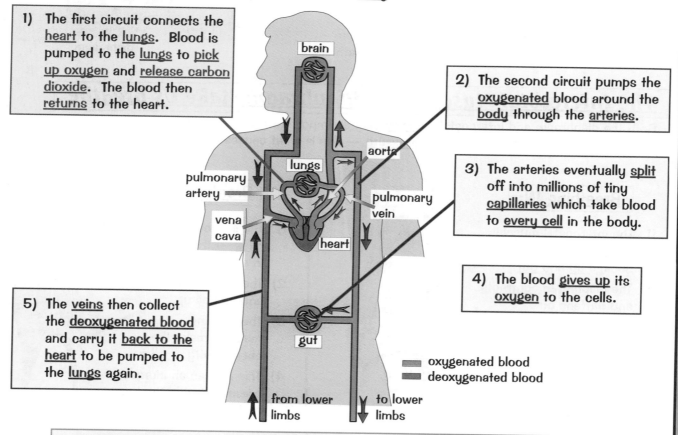

1) The first circuit connects the heart to the lungs. Blood is pumped to the lungs to pick up oxygen and release carbon dioxide. The blood then returns to the heart.

2) The second circuit pumps the oxygenated blood around the body through the arteries.

3) The arteries eventually split off into millions of tiny capillaries which take blood to every cell in the body.

4) The blood gives up its oxygen to the cells.

5) The veins then collect the deoxygenated blood and carry it back to the heart to be pumped to the lungs again.

brain

aorta

lungs

pulmonary artery

pulmonary vein

vena cava

heart

gut

oxygenated blood
deoxygenated blood

from lower limbs

to lower limbs

As the blood passes through the gut it picks up glucose. As it travels around the rest of the body, it drops off the glucose where it's needed for respiration (see page 66).

Okay — let's get to the heart of the matter...

The human heart beats 100 000 times a day on average. You can feel a pulse in your wrist or neck (where the arteries are close to the surface). This is the blood being pushed along by a heartbeat.

The Lungs and Breathing

You need to get air (containing <u>oxygen</u>) into your lungs so the oxygen can get into the blood... which is where <u>breathing</u> comes in. (They sometimes call it '<u>ventilation</u>' in the exams — don't get confused with those big shiny metal things that Bruce Willis likes climbing though, it's just breathing, OK.)

The Thorax — The Top Part of Your Body

The <u>thorax</u> is the part of the 'body' from the neck down to the diaphragm. There are a few parts you need to know...

1) The <u>lungs</u> are like big pink <u>sponges</u>.

2) The <u>trachea</u> (the pipe connecting your mouth and nose to your lungs) splits into two tubes called '<u>bronchi</u>' — one goes to each lung.

3) The bronchi split into progressively smaller tubes called <u>bronchioles</u> that end with small bags called <u>alveoli</u> — this is where oxygen moves into the blood and carbon dioxide moves out.

4) The <u>ribs</u> protect the lungs and the heart etc. They're also important in breathing (see below).

5) The <u>intercostal muscles</u> are the muscles in between the ribs.

6) The <u>diaphragm</u> is the large muscle at the bottom of the thorax, which is also important for breathing.

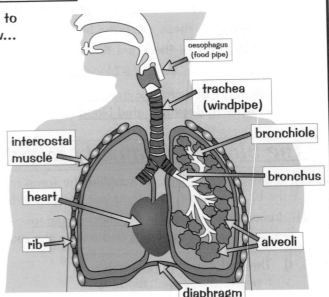

The Intercostal Muscles and Diaphragm Make Us Breathe

Both the <u>diaphragm</u> and <u>intercostal muscles</u> play an important role in breathing in and out. During ventilation there is a change in <u>pressure</u> — this is what causes air to enter and leave the lungs.

Breathing IN:

1) The intercostal muscles <u>contract</u>, pulling the ribcage <u>up</u> and <u>out</u>.

2) The diaphragm <u>contracts</u> and flattens out.

3) The contracting muscles make the chest cavity <u>larger</u>.

4) This <u>decreases</u> the <u>pressure</u> inside the lungs, so air is drawn <u>in</u>.

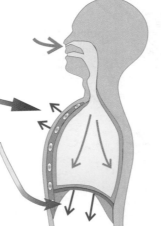

Breathing OUT:

1) The intercostal muscles <u>relax</u> — the ribcage drops in and down.

2) The diaphragm <u>relaxes</u> and arches up.

3) The relaxing muscles make the chest cavity <u>smaller</u>.

4) This causes an increase in pressure inside the lungs, and air is forced <u>out</u>.

Take a breather — there's a lung way to go yet...

If you've ever fancied a career as a <u>sports physiologist</u> than you'll need to know this stuff inside out. Plus it comes in really handy in everyday life — I regularly drop interesting biology facts into conversation in an attempt to woo the opposite sex. Even if it doesn't go down too well, at least it's stuck in your head that air is drawn <u>into the lungs</u> because of a <u>decrease in pressure</u> caused by an <u>increased thorax volume</u>. Oh, and it'll also be useful in the exam. So many reasons to learn this page.

Measuring Lung Capacity

Get out your rulers, folks, because it's time to measure your lungs. (I lied about the ruler, though.)

Lung Capacity is Measured Using a Spirometer

The best way to measure lung capacity is with a device called a spirometer. It measures the volume of air breathed in and out, and you can use it to produce graphs like the one below. It can also help detect lung problems, e.g. asthma.

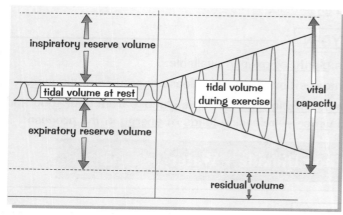

1) **TIDAL VOLUME** is the amount you breathe in (or out) with each breath.

2) **VITAL CAPACITY** is the most air you could possibly breathe in or out in one breath.

3) **INSPIRATORY RESERVE VOLUME** is the most air you could breathe in after breathing in normally.

4) **EXPIRATORY RESERVE VOLUME** is the most air you could force out after breathing out normally.

5) **RESIDUAL VOLUME** is the amount of air left in your lungs after you've breathed out as much as possible. This stops your lungs from collapsing.

MEASURING YOUR LUNG CAPACITY

1) To measure your tidal volume, breathe normally into the spirometer 3 times.

2) Divide your reading by 3 to get an average tidal volume.

3) To measure your vital capacity put the spirometer in your mouth.

4) Breathe in as far as you can through your nose and then out as far as you can through your mouth.

5) Repeat 2 more times then use the highest value.

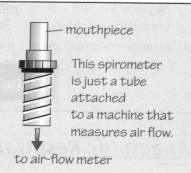

mouthpiece

This spirometer is just a tube attached to a machine that measures air flow.

to air-flow meter

Exercise Affects Tidal Volume

When you start to exercise, your breathing rate increases, and your lung capacity changes:

1) Your tidal volume increases — you take deeper breaths.

2) Your inspiratory and expiratory reserve volumes decrease.

3) Some athletes have a higher than average vital capacity, and a smaller residual volume, but in general there's not much difference in lung capacity between athletes and 'normal' people.

Now take a deep breath and learn these facts...

If you're asthmatic, you might be used to breathing into a type of spirometer called a peak flow meter. These measure your lung capacity and how much force you're using to breathe out. This helps doctors adjust the amount of medication you need to carry on with your active life. Hurrah.

Respiration and Exercise

Respiration might not sound very rock 'n' roll, but it's pretty fundamental to life as we know it.

Respiration is NOT 'Breathing In and Out'

Respiration is really important — it releases the energy that cells need to do just about everything.

1) Respiration is the process of breaking down glucose to release energy.

2) It goes on in every cell in your body. When you exercise, a lot of the energy released in respiration is used to make your muscles contract.

3) There are two types of respiration — aerobic and anaerobic.

AEROBIC RESPIRATION NEEDS PLENTY OF OXYGEN

1) Aerobic respiration is what happens when there's plenty of oxygen available.

2) "Aerobic" just means "with oxygen" and it's the most efficient way to release energy from glucose.

3) This is the type of respiration that you're using most of the time. It turns glucose from your food and oxygen from your lungs into carbon dioxide and water — releasing loads of energy in the process:

$$\text{Glucose + Oxygen} \longrightarrow \text{Carbon Dioxide + Water} \ (+ \text{ENERGY})$$

ANAEROBIC RESPIRATION DOESN'T USE OXYGEN AT ALL

1) Anaerobic respiration happens when there's not enough oxygen available.

2) "Anaerobic" just means "without oxygen". It's NOT the best way to convert glucose into energy because it releases much less energy than aerobic respiration.

3) In anaerobic respiration, the glucose is only partially broken down, and lactic acid is produced.

$$\text{Glucose} \longrightarrow \text{Lactic Acid} \ (+ \text{ENERGY})$$

Anaerobic Respiration Gives You an Oxygen Debt

Anaerobic respiration is okay for a short period of time, but it has some unpleasant side effects:

1) When you respire anaerobically lactic acid builds up in the muscles, which can be painful. Your muscles can become fatigued, and stop being able to respond — leaving you feeling weak and shaky.

2) The advantage is that at least you can keep on using your muscles for longer.

3) After resorting to anaerobic respiration, when you stop exercising you'll have an oxygen debt — your muscles are still short of oxygen because they haven't been getting enough for a while. You'll need extra oxygen to break down all the lactic acid that's built up in your muscles.

4) This means you have to keep breathing hard for a while after you stop exercising — to repay the debt.

5) Your heart rate has to stay high for a while too — to get the extra oxygen to your muscles.

6) Anaerobic respiration doesn't release much energy compared to aerobic respiration — but it's useful in emergencies. Unfit people have to resort to anaerobic respiration quicker than fit people do.

I reckon aerobics classes should be called anaerobics instead...

OK, so when you're just sitting about, you use aerobic respiration to get all your energy — but when you do strenuous exercise, you can't get enough oxygen to your muscles, so you use anaerobic respiration too. Nothing too taxing here — just make sure you know what an oxygen debt is.

Measuring Heart and Breathing Rate

Sports physiologists often <u>monitor</u> an athlete's <u>heart</u> and <u>breathing</u> rates before, during and after exercise. This can show how <u>fit</u> an athlete is, and what sort of <u>training programme</u> they need.

Exercising Increases Breathing Rate and Heart Rate

1) Muscles need <u>energy</u> from respiration to <u>contract</u>. When you exercise some of your muscles contract more frequently than normal so you need <u>more energy</u>. This energy comes from <u>increased respiration</u>.

2) The increase in respiration means you need to get <u>more oxygen</u> into the cells.

3) Your breathing rate increases to try and get <u>more oxygen into</u> the blood and <u>more CO_2 out</u> of the blood. Your <u>heart rate</u> increases to deliver <u>oxygen</u> and <u>glucose</u> to the <u>muscles</u> more <u>quickly</u>.

4) When you do <u>really vigorous exercise</u> (like sprinting) your body can't supply <u>oxygen</u> to your muscles quickly enough, so they start <u>respiring anaerobically</u> (see previous page).

Measure Heart Rate by Taking a Pulse

1) It's hard to measure heart rate directly, but you can measure <u>pulse rate</u> instead.

2) You take someone's pulse by <u>placing two fingers</u> on their <u>wrist</u> or neck and <u>counting</u> the <u>number of pulses you feel</u> in a <u>minute</u>.

3) Each pulse is <u>one</u> heartbeat.

4) Measure your pulse when you're sitting down quietly to find your <u>resting heart rate</u>. Most people have a resting heart rate of around <u>70 beats/min</u>, but it depends on how fit you are — <u>fitter</u> people have a <u>slower</u> resting heart rate.

> To measure someone's <u>breathing rate</u>, count the <u>number</u> of breaths they take per <u>minute</u>.

Recovery Time Depends on Fitness

1) Your <u>recovery time</u> is the time it takes your body to get back to normal.

2) It depends on how <u>strenuous</u> the exercise was, and how <u>fit</u> you are.

YOU CAN <u>MEASURE</u> YOUR RECOVERY TIME IN AN EXPERIMENT

1) Measure your <u>pulse rate</u> at <u>rest</u>.

2) <u>Run</u> about for <u>3 minutes</u>.

3) Measure your pulse rate <u>every two minutes</u> till it's back to <u>normal</u>.

4) The <u>time</u> it takes from when you <u>stop</u> exercising for your pulse to go back to <u>normal</u> is your <u>recovery time</u>.

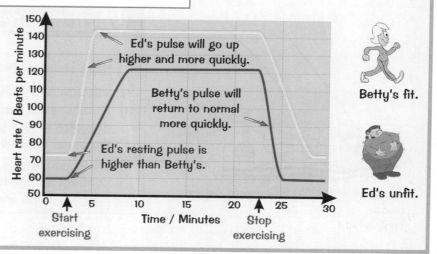

All this talk of exercise is making me feel tired...

Many athletes these days use an <u>electronic</u> heart rate monitor that straps onto their wrist or chest. This means they can measure their heart rate <u>while</u> they're exercising, then <u>download</u> the data onto a computer and analyse it. It's quite tricky taking your pulse when you're in the middle of a marathon.

Controlling Body Temperature

It's a frosty morning and you're outside shivering in your tiny shorts while the sadistic PE teacher yells at you to run faster. Five minutes later you're pouring with sweat, with a face like a beetroot — just great.

Body Temperature is Around 37 °C

1) All reactions in your body (e.g. respiration — see p.66) work best at about 37 °C.
2) This means that you need to keep your body temperature around this value — within 1 or 2 °C of it.
3) If the body gets too hot or too cold, some really important reactions could be disrupted. In extreme cases, this can even lead to death.

Your Body Has Some Tricks for Altering Body Temperature...

1) There is a thermoregulatory centre in the brain which acts as your own personal thermostat.
2) It contains receptors that are sensitive to the temperature of the blood flowing through the brain.
3) The thermoregulatory centre also receives impulses from the skin, giving info about skin temperature.
4) If you're getting too hot or too cold, your body can respond to try and cool you down or warm you up:

When you're too HOT:

1) Hairs lie flat.
2) Lots of sweat is produced by sweat glands and evaporates from the skin, which removes heat.
3) The blood vessels supplying the skin get bigger in diameter — so more warm blood flows close to the surface of the skin and more heat is lost to the surroundings.

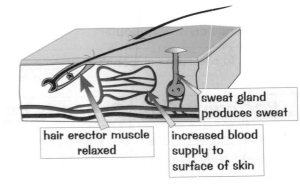

hair erector muscle relaxed

sweat gland produces sweat

increased blood supply to surface of skin

When you're too COLD:

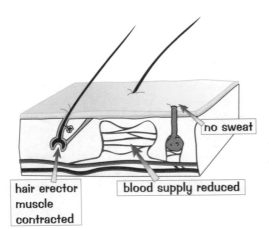

hair erector muscle contracted

blood supply reduced

no sweat

1) Hairs stand up to trap an insulating layer of air.
2) Less sweat is produced.
3) Blood vessels supplying skin capillaries get smaller in diameter to reduce the skin's blood supply. This stops the blood losing so much heat to the surroundings.
4) When you're cold you shiver too (your muscles contract automatically). This needs respiration, which releases some energy as heat.

Sweaty and red — I'm so attractive in the heat...

If you get really hot you could get heat exhaustion and if it's not treated, you could die — scary. Same thing if you get too cold (hypothermia) — you can go into a coma and die. If your sticking-out bits (like fingers) get too cold they can get frostbite — they go black and might have to be cut off. Ew.

Controlling Water and Glucose

Your body has to work quite <u>hard</u> to keep the levels of all the things that pass in and out nice and <u>stable</u>. This page is about how your body regulates two <u>really important</u> substances — <u>water</u> and <u>glucose</u>.

We Need to Balance Our Water Content

The body <u>constantly</u> has to <u>balance</u> the water coming in against the water going out.

Water is taken into the body as <u>food and drink</u> and is <u>lost</u> from the body in <u>three main ways</u>:

1) In <u>urine</u>   2) In <u>sweat</u>   3) In the air we <u>breathe out</u>.

Our bodies can't control how much we lose in our breath, but we do control the other factors. This means the <u>water balance</u> depends on:

1) the liquids <u>consumed</u>

2) the amount <u>sweated out</u>

3) the amount <u>excreted by the kidneys</u> in the <u>urine</u>.

On a <u>cold</u> day, if you <u>don't sweat</u>, you'll produce <u>more urine</u>, which will be <u>pale</u> and <u>dilute</u>.

On a <u>hot</u> day you <u>sweat a lot</u>, and you'll produce <u>less urine</u>, which will be <u>dark-coloured</u> and <u>concentrated</u>.

The water lost when it's hot has to be <u>replaced</u> with water from food and drink to restore the <u>balance</u>.

Insulin and Glucagon Control Blood Sugar Levels

1) Eating foods rich in <u>carbohydrate</u> puts a lot of <u>glucose</u> into the blood from the <u>gut</u>.

2) Normal <u>respiration</u> (see p.66) in cells <u>removes</u> glucose from the blood.

3) Vigorous <u>exercise</u> also removes a lot of glucose from the blood.

4) Levels of glucose in the blood must be kept <u>steady</u>. <u>Changes</u> in blood glucose are monitored and controlled by the <u>pancreas</u>, using the hormones <u>insulin</u> and <u>glucagon</u>...

Blood glucose level TOO HIGH — insulin is ADDED

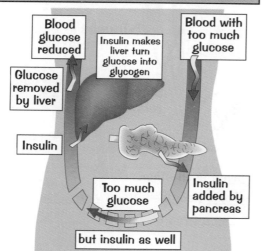

Blood glucose level TOO LOW — glucagon is ADDED

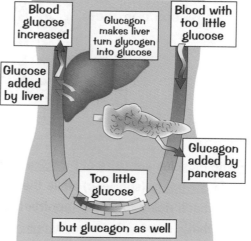

<u>Glycogen</u> can be stored in the <u>liver</u> until blood sugar levels get low again.

My glucose levels are falling — pass the biscuits...

If you're exercising hard, make sure you <u>drink</u> enough water, to make up for what you're losing in <u>sweat</u>. Getting <u>dehydrated</u> can be really unpleasant and <u>dangerous</u>, and you won't be able to do your best in any sport if you're short of water. Don't drink too much water, though, cos that's bad too... Hmmm, tricky.

The Action of Muscles

You need to know about how muscles work in pairs, and all the fancy names that are used to describe this. Once you've got the hang of it you should have a pretty good idea how muscles move your bones around.

Muscles Pull on Bones

Each muscle is attached to at least two different bones by tendons.
Only one of these bones will move when the muscle contracts.

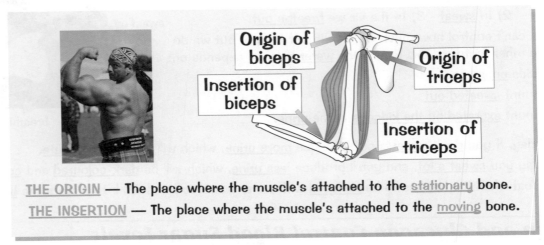

Origin of biceps
Insertion of biceps
Origin of triceps
Insertion of triceps

THE ORIGIN — The place where the muscle's attached to the stationary bone.
THE INSERTION — The place where the muscle's attached to the moving bone.

Antagonistic Muscles Work in Pairs

Muscles can only do one thing — pull. To make a joint move in two directions, you need two muscles that can pull in opposite directions.

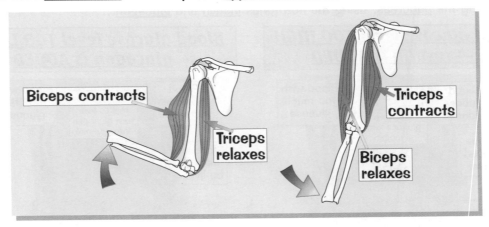

Biceps contracts
Triceps relaxes
Triceps contracts
Biceps relaxes

1) Antagonistic muscles are pairs of muscles that work against each other.
2) One muscle contracts (shortens) while the other one relaxes (lengthens) and vice versa.
3) The muscle that's doing the work (contracting) is the prime mover, or agonist.
4) The muscle that's relaxing is the antagonist.

Tricepatops — it's an 'armless dinosaur...

Muscles aren't too tricky as long as you remember a couple of key things — they move your bones, and if they're antagonistic, they work in pairs. Try moving your arm just like in these diagrams — you can see the muscles contracting. You might look like a bit of an idiot if people are watching, though.

Measuring Glucose and Muscle Strength

You need to know how to carry out the following tests for the <u>exam</u>, but no one's going to ask you to wee on a stick or anything. <u>Don't</u> go testing your blood at home either — it wouldn't be a great idea.

Glucose Levels are Measured Using Dipsticks

You can test <u>urine</u> and <u>blood</u> to measure <u>glucose levels</u>. People with <u>diabetes</u> (a disease where you don't produce enough <u>insulin</u>) have to measure their blood glucose levels <u>regularly</u>.

TESTING URINE

1) Dip a <u>glucose test strip</u> (dipstick) into a urine sample.
2) Compare it against the manufacturer's <u>colour chart</u> — this shows if glucose is present.
3) There <u>shouldn't be any glucose</u> in your urine — a <u>positive result</u> might mean that you have <u>diabetes</u>.

CORDELIA MOLLOY / SCIENCE PHOTO LIBRARY

TESTING BLOOD

SATURN STILLS / SCIENCE PHOTO LIBRARY

1) <u>Prick</u> your finger using a special sterile 'pen'.
2) Put a <u>drop</u> of blood onto a <u>blood glucose test strip</u>.
3) Compare it to the manufacturer's <u>colour chart</u>.
4) The strip turns a different colour depending on how much glucose is present.
5) This helps a <u>diabetic</u> work out how much <u>insulin</u> to inject.
6) You can also get <u>digital</u> blood glucose meters that give a more <u>accurate</u> reading.

Muscle Strength is Measured Using the Grip Test

You can test the <u>strength</u> of your <u>forearm</u> and <u>hand</u> muscles using a <u>handgrip dynamometer</u>. This is sometimes used as a measure of how much <u>overall muscle strength</u> you have.

1) <u>Adjust</u> the dynamometer to match your hand size.
2) Hold the handgrip dynamometer in your <u>right hand</u> (if you're right-handed) in line with your forearm, with your elbow at a right angle to your body.
3) <u>Grip</u> the dynamometer as hard as you can. <u>Repeat</u> 3 times.
4) Record the <u>best</u> reading (usually in <u>kg</u>). <u>Average</u> strengths are about 26-31 kg for women, and 46-51 kg for men, but it varies a lot.

You can <u>repeat</u> this test over a period of strength training, to monitor your <u>progress</u>.

The handgrip test is also used to assess <u>wrist</u> and <u>hand</u> <u>injuries</u>, and how they're responding to <u>treatment</u>.

If you don't know these tests, urine big trouble...

Some people use the <u>grip test</u> to show <u>overall muscle strength</u>. That isn't always <u>reliable</u>, because it only tests your hand and forearm muscles. People in <u>certain jobs</u> usually have a strong handgrip, e.g. <u>plumbers</u>, but it <u>doesn't</u> mean that the rest of their muscles are as strong as Superman on steroids...

Revision Summary for Section 2.7

Who would have guessed that studying sport would involve so much biology — I was under the impression you just needed to run about a bit and maybe catch or hit some balls. But as any good sportsperson will tell you, it's important to know about what happens in your body during exercise, so you can improve your fitness level. Here are some questions to test that you know your stuff. Make sure you can answer all of them before moving on.

1) Name five things that success in sport depends upon.

2) What is cardiovascular fitness?

3) Name the three types of blood vessel and describe what each one does.

4) Do arteries normally carry oxygenated or deoxygenated blood?

5) What organ pumps blood around the body?

6) Humans have a double circulatory system. What does this mean?

7) Describe the movement of blood around the circulatory system.

8) Where in the body does the blood pick up oxygen?

9) What is the thorax?

10) Put the following parts of the respiratory system into the correct order (from the mouth to the lungs): a) alveoli, b) trachea, c) bronchiole, d) bronchus.

11) Describe what happens in the chest to cause you to: a) breathe in, b) breathe out.

12) What would a doctor use to measure lung capacity? Why would you want to measure it?

13) What do the following terms mean? a) tidal volume, b) expiratory reserve volume, c) residual volume.

14) What affect does exercise have on a person's tidal volume?

15) What is respiration?

16) What is the difference between aerobic and anaerobic respiration? Give the word equation for each.

17) Describe the advantages and disadvantages of anaerobic respiration.

18) Why does an 'oxygen debt' develop in muscles that are respiring anaerobically?

19) How does exercising effect a person's breathing rate and heart rate?

20) How would you measure someone's heart rate?

21) How would you measure someone's breathing rate?

22) Describe how you could monitor recovery time.

23) What temperature is the human body kept at?

24) How does the body reduce its temperature when it's too hot?

25) How does the body increase its temperature when it's too cold?

26) Give three ways in which we lose water.

27) Describe how insulin and glucagon control blood sugar levels.

28) How are muscles attached to bones?

29) Describe what antagonistic muscles are, using the biceps and triceps as an example.

30) Describe how you could measure the glucose level in: a) urine, b) blood.

31) How could you measure muscle strength?

Finding the Optimum Diet

To perform at their best, athletes need to be healthy — and that means eating a <u>healthy diet</u> (see p.12).

Athletes <u>Need Different Diets</u> from Other People

Because athletes push their bodies harder than most of us, they have different <u>dietary requirements</u>:

1) They need plenty of <u>carbohydrates</u> to provide the <u>energy</u> for training and competition.
2) They might need extra <u>protein</u> to <u>build</u> and <u>repair muscle</u>.
3) They need to drink more <u>water</u> to stay <u>hydrated</u> and regulate their <u>body temperature</u>.
4) They need to monitor their <u>fat</u> intake to ensure it stays within <u>healthy</u> limits.

And different types of athletes need <u>different amounts</u> of each nutrient.
For example:

	Carbohydrate (g/day)	Protein (g/day)	Energy (Kcal/day)
Average non-athlete	375	70	2500
Bodybuilder	600	180	4000
Marathon runner	1000	150	6000

Sports Nutritionists <u>and Dieticians</u> Study Athlete's Diets

<u>Sports nutritionists</u> and <u>dieticians</u> provide <u>nutritional advice</u> to athletes. They work out the best diet for athletes so they can stay <u>healthy</u> and <u>optimise their performance</u>. By studying what the athlete currently eats, they can also work out if they are getting <u>everything</u> they need in the <u>right amounts</u>, and give <u>advice</u> about how to improve their diet.

They may ask the athlete to do a '<u>24-hour dietary recall</u>' or a '<u>diet diary</u>'.

1) '<u>24-hour dietary recall</u>' — the athlete is asked to write down <u>everything</u> they ate and drank in the last <u>24 hours</u>. This is a quick easy method for the nutritionist to get an idea of the athlete's <u>eating habits</u>.

2) '<u>Diet diaries</u>' — the athlete is asked to write down <u>everything</u> they eat and drink for a week or even longer. The nutritionist can examine the diary, work out the athlete's <u>nutrient intake</u>, and provide <u>advice</u> on how to improve the athlete's diet and performance.

Here is a '24-hour dietary recall' from a <u>weightlifter</u>, complete with a nutritionist's comments:

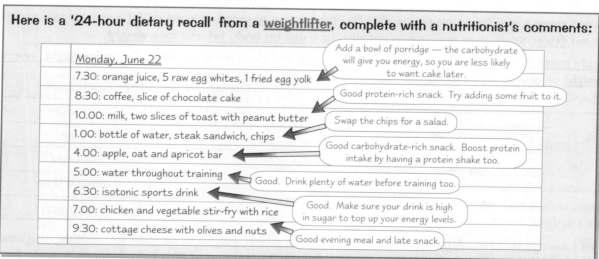

<u>Monday, June 22</u>
7.30: orange juice, 5 raw egg whites, 1 fried egg yolk — Add a bowl of porridge — the carbohydrate will give you energy, so you are less likely to want cake later.
8.30: coffee, slice of chocolate cake — Good protein-rich snack. Try adding some fruit to it.
10.00: milk, two slices of toast with peanut butter — Swap the chips for a salad.
1.00: bottle of water, steak sandwich, chips — Good carbohydrate-rich snack. Boost protein intake by having a protein shake too.
4.00: apple, oat and apricot bar
5.00: water throughout training — Good. Drink plenty of water before training too.
6.30: isotonic sports drink — Good. Make sure your drink is high in sugar to top up your energy levels.
7.00: chicken and vegetable stir-fry with rice
9.30: cottage cheese with olives and nuts — Good evening meal and late snack.

Dear diet diary, today I have devoured...

Athletes have to be very careful with what they eat <u>every day</u>. So there's no snacking on chocolate, or gorging on ice cream if you've had a bad day, and definitely no pork scratchings.

Energy Needs

You need energy <u>all the time</u>, even when sleeping. How much someone needs depends on the <u>individual</u>.

Different People **Need** Different Amounts **of Energy**

The <u>amount of energy</u> you need depends on your <u>body mass</u> and your level of <u>activity</u>, so the <u>heavier</u> and the <u>more active</u> you are, the <u>more energy</u> you will need.

A LARGER BODY MASS NEEDS MORE ENERGY

Every <u>cell</u> in the body needs <u>energy</u>. So the <u>bigger</u> you are, the <u>more cells</u> you have, and the more energy you'll need. You also need <u>energy</u> to <u>move</u> your body, and it takes <u>more</u> energy to move a <u>bigger mass</u>.

You can work out how much energy you need:
For every <u>kg</u> of <u>body mass</u>, you need <u>1.3 Kcal</u> every <u>hour</u>.
This is the <u>basic energy requirement</u> (BER) needed to maintain <u>essential</u> body functions.

> daily basic energy requirement (Kcal/day) = 1.3 × 24 hours × body mass (kg)
> E.g. a person weighing 60 kg would require 1.3 × 24 × 60 = 1872 Kcal/day

YOU NEED MORE ENERGY IF YOU EXERCISE

The <u>more active</u> you are the <u>more energy</u> you will need.
For example, in half an hour a person who weighs 60 kg will use (on average):

1) 100 Kcal <u>walking</u>   2) 180 Kcal <u>cycling</u>   3) 350 Kcal <u>running</u>

- An <u>average person</u> needs 2000–3000 Kcal a day.
- <u>Bodybuilders</u> often eat up to 5000 Kcal a day when 'bulking up' for competitions.
- The <u>cyclist</u> Lance Armstrong burnt 6000-7000 Kcal a day when cycling in the Tour de France.

Body Mass Index **Indicates If You're Under- or Overweight**

If you eat <u>more</u> Kcals than you <u>use</u> in activity, the <u>excess energy</u> is stored as <u>fat</u> and you gain weight.
If you eat <u>fewer</u> Kcals than you <u>use</u> in activity, you <u>use up body fat</u> and <u>lose weight</u>.

The <u>body mass index</u> (BMI) is used as a guide to help decide whether someone is <u>underweight</u>, <u>normal</u>, <u>overweight</u> or <u>obese</u>. It's calculated from their <u>height</u> and <u>weight</u>:

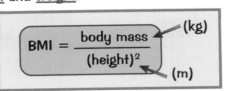

$$BMI = \frac{body\ mass\ \text{(kg)}}{(height)^2\ \text{(m)}}$$

Body Mass Index	Weight Description
below 18.5	underweight
18.5 - 24.9	normal
25 - 29.9	overweight
30 - 40	moderately obese
above 40	severely obese

The table shows how BMI is used to <u>classify</u> people's weight.

BMI isn't always reliable. <u>Athletes</u> have lots of muscle, which weighs more than fat, so they can come out with a high BMI even though they're not overweight. An alternative to BMI is measuring <u>% body fat</u>.

You use 65 Kcal an hour just sleeping...

Isn't that wonderful to know (but of course, you use up more if you're actually doing something).

Energy-Rich and Muscle-Building Diets

There are a few things athletes can do to <u>improve</u> their <u>performance</u>... and no, I don't mean by taking steroids, or growth hormones, or anything else that's highly illegal — I mean by changing their diet.

Athletes Eat More Carbohydrates Before Competitions

A few days before a competition, <u>endurance athletes</u> (like marathon runners and long-distance cyclists) often <u>increase</u> their intake of <u>complex carbohydrates</u>, e.g. by eating more pasta, bread and rice. This is called '<u>carbohydrate loading</u>'.

1) <u>Carbohydrates</u> are broken down in the body to release <u>glucose</u>.
2) <u>Excess glucose</u> can be <u>stored</u> in the <u>muscles</u> as a chemical called <u>glycogen</u>.
3) When the muscles need more energy, <u>glycogen</u> is converted back into <u>glucose</u>, which is broken down to release <u>energy</u> (see p.66).
4) '<u>Carbohydrate loading</u>' increases the amount of glycogen stored in their muscles, so when athletes are running long races, they can run for longer without getting tired.

Carbohydrate loading can have some adverse side effects like <u>muscle stiffness</u> and <u>chest pains</u> — so athletes are advised to only do it a few times a year.

EXAMPLE: TRAINING FOR A LONG-DISTANCE RACE

Nikki has just starting <u>training</u> for the Great North Run. She's changed her diet to <u>prepare</u> for competition by:

1) Making sure she eats a <u>balanced diet</u>, including lots of fruit and vegetables.
2) Increasing her intake of <u>complex carbohydrates</u>.
3) Eating more <u>snacks</u> — to provide more <u>energy</u>.
4) Making sure she drinks enough <u>water</u>, especially before she goes out training.

Nikki's Daily Diet
8.00: bowl of cereal, banana, cup of tea
11.00: two slices of toast and butter
1.00: pasta with fish and vegetables, fruit, yoghurt
5.00: two slices of bread and butter, banana
6.30: half a litre of water
8.00: potato omelette, fruit
(Water drunk throughout the day.)

Some Athletes Need a High Protein Diet

1) <u>Protein</u> is needed by the body to <u>build</u> and <u>repair muscle</u>.
2) Some athletes, like <u>bodybuilders</u> and <u>powerlifters</u>, eat a <u>high</u> protein diet to help them <u>increase</u> their <u>muscle mass</u>.
3) This involves eating lots of <u>protein-rich foods</u>, like meat, eggs and pulses.
4) They may also drink <u>protein shakes</u> — these are drinks rich in protein that you can buy or make at home. The ingredients can include milk, oats, whey, peanut butter, fish oil and raw eggs.

Information loading — only when exams are on...

This is where a good <u>nutritionist</u> can really come in handy. They know which foods contain the <u>right proportions</u> of the nutrients that you need, and they can tell you where you're going wrong. It's like my mum always says — you only get out as good as you put in...

Isotonic Sports Drinks

Imagine you've just run a marathon. You're exhausted and thirsty. You reach for a water bottle...
Think again, and swap the water for an <u>isotonic sports drink</u>. Here's why.

During Exercise You Lose Water, Glucose and Electrolytes

During exercise your body <u>respires more</u>. This means:

1) You use up a lot of <u>glucose</u> by breaking it down to release <u>energy</u>.

2) You get <u>warmer</u> — which means you <u>sweat</u> to <u>cool down</u>.

3) Sweat contains lots of <u>water</u> and ions called <u>electrolytes</u> — so you're losing these from your body.

After exercise you need to <u>replenish</u> your body's supplies.

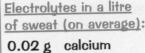

<u>Electrolytes in a litre of sweat (on average):</u>

0.02 g calcium
0.05 g magnesium
1.15 g sodium
0.23 g potassium
1.48 g chloride

Isotonic Drinks Contain Water, Glucose and Electrolytes

1) Isotonic sports drinks are designed to help athletes <u>recover</u> after exercise.

2) They contain <u>water</u>, <u>glucose</u> and <u>electrolytes</u> in the <u>same concentrations</u> as they are in the body.

3) So they <u>replace</u> the water and electrolytes lost during sweating, and top up carbohydrate stores.

Many athletes use isotonic sports drinks to rehydrate after <u>strenuous training</u> and <u>competition</u>. <u>Several brands</u> of isotonic sports drinks are available to buy.

WARNING: Because isotonic sports drinks contain high levels of glucose (a sugar), they are bad for your teeth and can cause weight gain if used by non-athletes. Sports drinks should only be used if you train to a fairly high level (3 – 5 times a week for 45 – 90 minutes).

<u>Sports drinks</u> are different from <u>energy drinks</u> — energy drinks just contain lots of <u>sugar</u> (and often <u>caffeine</u>).

Isotonic Drinks Rehydrate You Better Than Water

Drinking an isotonic sports drink is better than <u>drinking water</u> if you are <u>dehydrated</u> because:

1) Drinking water can make you <u>bloated</u> and <u>suppress your thirst</u> — this can put you off drinking enough to replace your losses.

2) If you <u>don't</u> have enough <u>ions</u> in your blood, you struggle to <u>retain</u> the water you drink — it takes a lot longer to be absorbed. Because isotonic drinks <u>replace</u> the ions lost during exercise, your body can retain the water in the drink.

3) Isotonic drinks replace the <u>electrolytes</u> you have lost, as well as replacing the <u>water</u>.

Ice 'n' tonic — I prefer my gin neat...

So not only do isotonic drinks give you bags of energy, they're <u>better</u> at replacing water than water is — who'd have thought it. Plus they always come in really good flavours — it's enough to make anybody want to train three to five times a week for 45 - 90 minutes.

Revision Summary for Section 2.8

Any serious athlete should know the stuff in this section — and so should you. For serious athletes it's all about having the right combination of nutrients to optimise their performance. For you, it's all about doing the right amount of reading, revision and testing yourself to pass your exam. And that's what this revision summary is for — the testing part at least.

1) Why do the dietary requirements for athletes differ from those for most of us?

2) Suggest why athletes need to monitor their fat intake more carefully than non-athletes.

3) What do sports nutritionists and dieticians do?

4) What is a '24-hour dietary recall'?

5) What is a 'diet diary'?

6) Look at the following page from a diet diary. Do you think this person is an athlete? Explain your answer.

Friday 29th Decemeber
8.00: bacon and egg sandwich, coffee
11.00: fruit
1.00: soup with bread, coffee
5.00: biscuits
7.00: lasagne with chips and salad, ice cream
(Squash drunk throughout the day.)

7) Why do you need more energy if you have a larger body mass?

8)* What is the daily basic energy requirement for a person weighing 75 kg?

9) Why do you need more energy if you're more active?

10)* Put these people in order of how much energy they are likely to need from their food (from highest to lowest), assuming that they all have the same mass:

   a) builder, b) professional runner, c) waitress, d) secretary.

11) What happens to you if you eat more Kcals than you need? What about fewer Kcals?

12) What is body mass index used for?

13)* Barry is an athlete who weighs 82 kg and has a height of 1.77 m. Calculate his BMI, and give his weight description. Suggest why this weight description might not be accurate.

14) What is 'carbohydrate loading'?

15) What is glucose stored as in the muscles?

16) Give three foods high in complex carbohydrates.

17) Athletes eat a lot of complex carbohydrates just before an event. Why shouldn't athletes eat the same amount of complex carbohydrates all year round?

18) What is protein needed for in the body?

19) Why do some athletes eat a high protein diet?

20) Give three foods high in protein.

21) What are protein shakes?

22) Give three ways a typical non-athlete can change their diet when training for a big race.

23) What do isotonic drinks contain?

24) Why do isotonic drinks rehydrate you better than water?

*Answers on p.100.

Designing Sportswear

Your equipment can make the difference between winning and losing — so the materials used to make sports clothing and equipment need to have just the right properties to do the job. It's the job of a materials scientist to design, create and test new materials with better and better properties.

Sports Clothing Has to be Light, Durable and Comfortable

1) If sports clothing is uncomfortable, it could be distracting and affect performance.

2) If it's too heavy, you need to put in extra effort to move around.

3) If it's not durable enough, it could be too expensive to keep replacing, and might not do the job properly — you don't want your motorbike trousers wearing out mid-skid.

Item	Used to be made of	Now made of	Because
Cycling helmets	Strips of leather	Expanded polystyrene (usually with a hard outer shell)	Leather rots when cyclists sweat. Expanded polystyrene is more durable and comfortable, and a better shock absorber.
Motorbike jacket	Leather	Nylon weave	Nylon weave is as strong and durable as leather, but lets air pass through it. It's lighter and more comfortable.

Friction Plays a Big Part in Design

1) Friction is the force that tries to stop two materials that are touching from moving past each other.

2) Air resistance is friction between a moving object and the air. It reduces speed, which is bad for fast sports where a fraction of a second can make all the difference.

3) Friction between materials isn't always a bad thing, though — it depends on the piece of equipment. If there were no friction between your shoes and the ground, you'd slip when you tried to walk.

4) Designers try to minimise or maximise friction to get the best out of equipment and sportswear, e.g.

Friction between	and	means that...	so we...
Running shoes	Running track	Runner is pushed forwards (a good thing)	Choose material and pattern of sole to increase friction.
Bicycle wheel	Brake pad	Wheel slows down (a good thing)	Choose brake pad material to increase friction and be durable.
Canoe	Water	Canoe is slowed down (a bad thing)	Make canoes a streamlined shape to reduce friction.
Canoe paddle	Water	Canoe is pushed forward (a good thing)	Increase the size of the paddle to increase friction.
Bicycle helmet	Air	Cyclist is slowed down (a bad thing)	Streamline helmets to reduce friction (air resistance).

Lightness, comfort and durability — you can't beat a mawashi...

For those of you who were wondering, a mawashi is one of those nappy-type things that sumo wrestlers wear — well ventilated, practical and stylish — if it was up to me we'd all wear them for work.

Materials for Equipment

When <u>designing</u> sports equipment one of the most <u>important</u> things is to choose the <u>right material</u> — you wouldn't get too far canoeing with a <u>canoe</u> made out of <u>concrete</u>.

Different Materials are Used for Different Equipment

1) Wood — often replaced by modern materials

Wood is <u>cheap</u>, fairly <u>strong</u>, <u>light</u> and <u>flexible</u>.

In the past wood was used for <u>skis</u>, <u>vaulting poles</u>, <u>rackets</u>, <u>surfboards</u>, etc.

Although it tends to have been replaced by more modern materials, wood is <u>still used</u> for some things, e.g. <u>cricket bats</u>, some <u>golf clubs</u> and <u>gymnastics beams</u>.

2) Metal — used for its stiffness and strength

Metals are very <u>stiff</u> and <u>strong</u>, but also <u>heavy</u>.

Metals are used to make <u>javelins</u>, <u>shots</u>, <u>bicycle frames</u>, <u>golf clubs</u>, <u>running spikes</u> and <u>fastenings</u> for clothes.

<u>Aluminium</u> is <u>stiff</u> enough for ski poles, and if it's made into <u>hollow tubes</u> it's still stiff, but much <u>lighter</u>.

3) Ceramics — not used for much in sports, really

Ceramics are materials that come from <u>minerals</u>. Clay and glass are good examples. They're <u>stiff</u> and <u>resist heat</u>, but are <u>heavy</u> and can <u>break easily</u>.

Ceramics are used in sports cars to make <u>brakes</u> and <u>mirrors</u>.

4) Polymers — have a wide range of properties

Atoms can join together to make <u>molecules</u>. Some small molecules can join together into <u>long chains</u> called <u>polymers</u>. All rubbers and plastics are polymers. They have lots of different properties, but they're mostly <u>light</u>, <u>flexible</u> and <u>waterproof</u>. Some polymers are <u>soft</u>, <u>squashy rubbers</u>, while others are <u>hard</u>, <u>stiff plastics</u>.

Polymers are used to make <u>artificial fibres</u> (e.g. polyester and nylon) for <u>clothes</u>, <u>rubber soles</u> for <u>shoes</u>, <u>bicycle tyres</u>, <u>balls</u> and <u>cycling helmets</u>. Polymers are not new — balls used to be made from natural rubber that came from the sap of a tree.

5) Composites — great but expensive

A composite material is made from <u>two or more other materials</u>, to <u>combine</u> their <u>properties</u>. Fibreglass is a good example — glass fibres are embedded in plastic to make a material that is <u>light</u> and <u>strong</u>.

Composites are used to make <u>surfboards</u>, <u>rackets</u>, <u>kayaks</u>, <u>skis</u> and <u>bicycle frames</u>.

Properties of Wood — he's got one in Milan, a flat in Paris...

Well, now you know what each material is <u>used for</u>. The next few pages describe in a bit more <u>detail</u> why they're used for that <u>particular activity</u>. They've all got <u>properties</u> that make them <u>especially good</u> at their job, a bit like me — I'm really good at spilling witch means I'm really good at bean a editor.

Metals and Ceramics

Metals and ceramics have <u>different properties</u> — these mean that they can be used for a <u>range</u> of different things, like swords and various different parts of sports cars. Cool.

Metals are Strong but Heavy

Not all metals have exactly the <u>same</u> properties, but in <u>general</u>:
1) They have a <u>high tensile strength</u> — they don't <u>break</u> easily when <u>stretched</u>.
2) They're <u>flexible</u> — they can be <u>bent</u> without breaking.
3) They're <u>hard</u> — they're hard to <u>dent</u> when pressure is applied.
4) They're good <u>conductors of electricity</u> and <u>heat</u>.

Metals Have Many Uses in Sport

1) Metals' <u>strength</u> and <u>hardness</u> make them good for <u>running spikes</u> and <u>javelin points</u> that press into other materials.
2) The <u>swords</u> used for fencing (foils) are metal — they're <u>flexible</u> and can <u>bend</u> without breaking.
3) Metals are used in <u>electrical systems</u> in cars because they <u>conduct electricity</u>.
4) Their <u>thermal conductivity</u> means they're also used in <u>car radiators</u>.

Ceramics are Hard but Break Easily

Ceramics also have a number of <u>common features</u>:
1) They break quite easily when <u>stretched</u> — they have <u>low tensile strength</u>.
2) They break when <u>bent</u> — they're <u>brittle</u>, not flexible.
3) They're very <u>hard</u>.
4) They're <u>poor conductors</u> of <u>electricity</u> and <u>heat</u> — they're <u>insulators</u>.
5) They have very <u>high melting points</u>.

Ceramics are Useful in Sports Cars

Because ceramics are so heavy and easy to break, they only have <u>a few applications</u> in sport.

CERAMIC BRAKES

When a car brakes, there's <u>friction</u> between the <u>brake pads</u> and the <u>brake discs</u>. This friction makes the discs and pads very <u>hot</u>.
Brake discs need to have a <u>high melting point</u> — if the discs started to melt they'd be <u>useless</u> for stopping the car.
The discs must also be <u>durable</u> so they don't wear down too <u>quickly</u>.
The high melting point and hardness of ceramics makes them ideal for the discs and pads.

SPARK PLUGS

Engines work by <u>exploding</u> a mixture of petrol and air in a confined space.
<u>Spark plugs</u> light the petrol and air mixture to cause the explosion, using <u>electricity</u>.
But the <u>contacts</u> in each plug need to be <u>separated</u> by an <u>electrical insulator</u>.
<u>Ceramics</u> are good insulators, and they don't melt in the hot engine.

Ceramics in sports cars — don't they break going round corners...

You need to learn the <u>properties</u> of metals and ceramics and how those properties make them <u>suited</u> to what they're used for. There could be a question in the exam which asks you to give <u>advantages</u> of, say, using <u>ceramic brake pads</u> in sports cars, or something along those lines — easy stuff, marks in the bag.

Polymers

Polymers are just loads of little molecules <u>joined together</u> in big, long <u>chains</u>.

Polymers Have Many Useful Properties...

Artificial polymers have a number of <u>useful properties</u>, which means they can be used to make a whole load of <u>sports equipment</u>. Some of these properties include:

1) <u>LOW DENSITY</u> — for equipment that needs to be <u>lightweight</u>.
2) <u>FLEXIBLE</u> — they can be used to make equipment that has to <u>bend</u> or <u>stretch</u>.
3) <u>LOW THERMAL CONDUCTIVITY</u> — polymers can be used to make clothing that keeps you <u>warm</u>.
4) <u>STRONG</u> — some polymers are strong and <u>tough</u>.
5) <u>EASILY MOULDED</u> — they can be used to manufacture equipment with almost <u>any shape</u>.
6) <u>INEXPENSIVE</u> — polymers are often <u>cheaper</u> than alternative materials.
7) <u>WATER-RESISTANT</u> — they don't <u>corrode</u>, <u>rot</u> or <u>rust</u>.
8) Chemicals can be added to polymers to change their <u>properties</u> — e.g. <u>PLASTICISERS</u> make polymers <u>more flexible</u>.

...Which Means They're Used to Make Loads of Stuff

Windscreens in racing cars need to be <u>lightweight</u> and <u>strong</u>. They're made from <u>high performance polymers</u> — which are also used to make visors and face shields for some sports.

Crash helmets need to be <u>strong</u>, <u>lightweight</u>, and <u>flexible</u>. The <u>hard</u> outer casing is made from a strong polymer and the inner lining is usually made from <u>polystyrene</u> foam, which is <u>flexible</u> and crushes on impact.

Kayaks need to be <u>lightweight</u> and <u>flexible</u> — so that they're fast and easy to manoeuvre. Many are made of polymers like <u>polythene</u>. Polythene kayaks don't <u>warp</u> or <u>rot</u> like traditional wooden ones.

Body armour for mountain biking or American football needs to be <u>strong</u>, <u>lightweight</u> and <u>flexible</u> — it's usually made of <u>polythene</u> too.

Flippers for scuba divers need to be <u>lightweight</u> and <u>flexible</u>.

Table tennis balls need to be <u>lightweight</u> — they're usually made out of a polymer called <u>celluloid</u>.

But there are some <u>DISADVANTAGES</u> to using polymers, too. Two of the biggest ones are:

1) They're made from <u>oil</u> — a <u>non-renewable resource</u>.
2) Most artificial polymers are <u>non-biodegradable</u>.

Plasticiser — sounds like a Schwarzenegger role...

It's a good job artificial polymers were invented, without them we'd have <u>rubbish kayaks</u>, <u>cycle helmets</u> and <u>ping pong balls</u>, but we'd also be without loads of other everyday things like <u>plastic bags</u> at the supermarket. But remember it's not all fun and games — there are also some <u>disadvantages</u>.

Composites

Composites are great, taking the good points of two different materials — it's a bit like a woolly jumper if you think about it — it keeps you warm and is really handy as a goal post but it doesn't eat grass all day.

Composites are Made of Two or More Different Materials

Composites are a mixture of two or more materials. The properties of a composite depend on the properties of the materials it is made from. For example:

1) Wood-plastic composites are made from wood and plastic. They behave like wood, and can be shaped using woodworking tools — but are water-resistant (like plastic).

2) Cermet is made from ceramic and metal. It's durable and heat-resistant (like ceramic), but malleable (like metal).

3) Carbon Fibre (or Carbon Fibre Reinforced Plastic — CRP) is made from a plastic (usually epoxy) and fibres made of carbon. It has a low density (like plastic) but is very strong (like carbon).

4) Fibreglass (or Glass Reinforced Plastic — GRP) is made from a plastic (usually polyester) and fibres made of glass. It has a low density (like plastic) but is very strong (like glass).

Glass Reinforced Plastic (GRP) is Strong and Lightweight

1) Glass Reinforced Plastic (GRP) is used to make things like surfboards, skis and many types of boat — including kayaks.

2) Like plastic, it's easily moulded into different shapes, lightweight, water-resistant and flexible.

3) The fibres of glass make it strong and durable.

Carbon Fibre is Even Lighter but More Expensive

Carbon Fibre Reinforced Plastic (CRP) is more expensive than Glass Reinforced Plastic, but it's even lighter. It's used to make the bodies of Formula 1 racing cars because it has a good strength-to-weight ratio — it's very strong (protecting the driver in a crash), and very light (enabling higher speeds).

Because CRP is so strong and light, it's used to make many other types of sporting equipment, including:

Bike frames — CRP is much lighter than traditional materials like aluminium or steel.

Fishing rods — CRP is lightweight and flexible, so you feel every bite.

Racket frames — CRP is strong enough to hold the strings in tension, and lightweight enough for easy handling.

Carbon fibre — it's CRP...

Carbon fibre comes in pretty handy because it's so strong and light. The problem is, it's also pretty pricey — you'd be lucky to get a carbon-fibre bike for less than a grand. It makes you think, just how much money is lost when two F1 cars crash into each other — and I thought my car insurance was bad.

Fabrics for Sportswear

Now you've got your <u>equipment sorted</u>, all you need to do is pick something to <u>wear</u>...

Fabrics Can be <u>Natural</u> or <u>Synthetic</u>

<u>Natural fabrics</u> are made from natural <u>plant</u> or <u>animal fibres</u>. For example:

1) <u>Cotton</u> is made from plant fibres. It's <u>soft</u>, <u>comfortable</u>, <u>breathable</u>, <u>strong</u> and <u>durable</u>. Undergarments and base layers are often made of cotton.

2) <u>Leather</u> is animal skin. It's <u>tough</u>, <u>strong</u>, <u>durable</u> and a good <u>thermal insulator</u>. Motorcyclists wear leather to <u>protect</u> them if they have an accident, and to keep them warm. But in sport it has been largely replaced by tough synthetic fabrics.

<u>Synthetic fabrics</u> are made from <u>artificial fibres</u>. For example:

1) <u>Polyester</u> is a type of polymer. It's <u>lightweight</u>, <u>durable</u> and <u>water-resistant</u>. Lots of sportswear is made of polyester — including football shirts, sports vests, fleecy jumpers and outdoor jackets.

2) <u>Elastane</u> (LYCRA®) is another type of polymer. It has excellent <u>elasticity</u>, it's <u>lightweight</u>, <u>durable</u> and <u>resistant to water</u> (and sweat). Sportswear which needs to be stretchy for <u>comfort</u> and fit is often made with elastane — for example swimwear, wetsuits, cycle shorts, athletic and aerobic clothing.

<u>Natural Fabrics are Comfy</u> but <u>Not So Good</u> If They Get Wet

Natural and synthetic fibres both have <u>advantages</u> and <u>disadvantages</u>, these can vary quite a lot but some <u>typical</u> advantages and disadvantages are shown in the table:

	Natural Fabrics	Synthetic Fabrics
Advantages	Very comfortable Breathable	Water-resistant Inexpensive
Disadvantages	Not water-resistant Heavy when wet	Can be uncomfortable Not breathable

These advantages and disadvantages <u>determine</u> a fabric's <u>uses</u>.

1) Natural fabrics are used when <u>comfort</u> is important — e.g. for underwear and base layers <u>next to the skin</u>. They're <u>unsuitable</u> as an <u>outer layer</u> in most outdoor sports — most of them don't perform well in the <u>wet</u>.

2) Synthetic fabrics (like polyester) are used to make <u>water-resistant</u> clothing for many <u>outdoor sports</u>. They are <u>unsuitable</u> to use for underwear or base layers, because they're not <u>breathable</u>.

Synthetic Fibres Can be Made More Comfortable

1) Synthetic fibres can be <u>mixed</u> with natural fibres to get the <u>best properties</u> of each. For example a <u>polyester-cotton mix</u> could be <u>comfortable</u> and <u>water-resistant</u>.

2) Special <u>polymer membranes</u> can be <u>bonded</u> to synthetic fabrics like polyester or nylon, to make them <u>breathable</u>. These fabrics are great for <u>outdoor clothing</u> — they're <u>waterproof</u> and <u>breathable</u> (so they allow sweat to escape, but don't let the rain in).

<u>Natural fibres — comfortable and they keep you regular...</u>

It's really important to get the <u>right clothing</u> for a sport — it'd be just plain daft to ride a <u>motorcycle</u> in <u>T-shirt and shorts</u> (yet they do it in Benidorm). And imagine playing <u>football</u> in something like <u>motorcycle gear</u> — you'd get all sweaty, though heading the ball might not hurt so much.

Choosing the Best Material

So many different materials to choose from — make sure you pick the right one for the job.

Choose Materials Based on Their Properties

Different properties of materials can be used to improve sports clothing and equipment in different ways.

Materials with a low thermal conductivity can help maintain your body temperature.

Smooth materials can be used to give equipment aerodynamic shapes.

Flexible materials can be used to make clothing and equipment that bends and stretches.

Shock-absorbing materials can be used to make equipment that withstands big forces.

Low-density materials can be used to make equipment lighter and faster.

Materials with a high tensile strength can be used to provide support or carry loads.

You Might be Asked to Pick the Right Material

In the exam they might ask you to pick a material to make a piece of sportswear or equipment. You need to pick the one with the best properties for the intended use.

Type of material	Properties
Aluminium	Strong and hard, lightweight for a metal
Steel	Strong, hard and heavy
Neoprene	Flexible, low thermal conductivity
Silk	Lightweight, soft and comfortable
Plasticised polymer	Flexible and light, can be easily moulded

Examples:

1) A fencing sword should be strong and hard enough to withstand clashes in a sword fight, and lightweight enough to be handled easily — aluminium is a good choice.

2) A scuba diving suit should be flexible to allow for comfortable movement, and provide enough insulation to keep divers warm — neoprene is a good choice.

3) The soles of tennis shoes should be flexible and lightweight to provide speed on court, and should be easily moulded into a shape that offers good grip so players can move without slipping — a plasticized polymer would be a good choice.

Statues of football legends — got to choose the Best material...

Think of the properties you need to make the equipment work well, then pick a material with exactly the same properties — not too difficult really, as long as you use a bit of common sense. Make sure you use all the stuff you can remember about the materials as well as any information they give you.

Revision Summary for Section 2.9

The thing I like most about this section is that it gives you a great excuse when you want some new trainers. Simply tell your loving mother that friction has become a bit of an issue — your current trainers don't provide enough grip and something a bit more lightweight and flexible would be nice — a plasticised polymer ought to do it. You could probably scam her out of a load of other stuff too, but make sure you know your stuff or you'll just look stupid — have a go at these, and see how you do.

1) Explain why sports clothing needs to be light, durable and comfortable.
2) Sports motorcycle jackets used to be made of leather. Why are they now often made of nylon weave?
3) State whether friction between the following is a good thing or a bad thing:
   a) Bicycle wheel and brake pad.   b) Bicycle helmet and air.
4) How does friction between running shoes and running tracks affect the design of running shoes?
5) Give one use of wood for sporting equipment.
6) Glass and clay are both examples of which type of material?
7) What is a polymer?
8) What features of metals makes them suitable for use as fencing swords?
9) Metals conduct heat. Give one way this feature can be used for sport.
10) List three properties of ceramics.
11) Give two uses of ceramics in sports cars.
12) What are the disadvantages of using polymers?
13) Give an advantage of making a kayak out of polythene instead of wood.
14) Why are crash helmets usually made from two different types of polymer?
15) When designing flippers for a scuba diver, what properties do they need to have?
16) Give one advantage of using a wood-plastic composite over just wood or just plastic.
17) What is GRP?
18) Why is carbon fibre used in the bodies of Formula 1 cars?
19) Give two other pieces of sporting equipment made from CRP.
20) Explain the difference between a natural fibre and a synthetic fibre. Give one example of each.
21) Give two advantages and two disadvantages of wearing natural fibres for sport.
22) Give a sport in which cotton sportswear would be unsuitable.
23) Give one example of a fabric which mixes natural fibres with synthetic fibres.
24)* Choosing from the materials in the table, which would you use for:
   a) Body armour for mountain biking?
   b) Clothing for riding a motorcycle?
   c) Making a high-performance bicycle?

Material	Properties
Carbon fibre	Low density, strong, lightweight, expensive
Polythene	Lightweight, hard, strong, water-resistant
Elastane	Elastic, lightweight, durable, flexible
Leather	Tough, strong, durable, flexible

* Answers on p.100.

The Roles of a Food Scientist

For Unit 3 you have to produce a <u>report</u> of a <u>practical investigation</u> that covers either <u>food science</u> or <u>forensic science</u> (see p.90) or <u>sports science</u> (see p.93).

Your Investigation Could be About Food Science

If you choose to study food science then your investigation should relate to food or the components found in food and food supplements. You'll need to:

- produce a <u>plan</u> and complete a <u>risk assessment</u> for the investigation.
- draw <u>conclusions</u> and <u>evaluate</u> your investigation.
- explain how a <u>food scientist</u> might use the <u>results</u> of your investigation and give a vocational application of your investigation.

Food Scientists Carry Out a Range of Different Roles

As part of your investigation you need to give a real-life, <u>vocational application</u> of your investigation — so first things first, you need to know what food scientists <u>actually do</u>...

1) Analyse Food Safety

Food scientists perform <u>laboratory tests</u> on food samples to find things like:

1) The <u>types</u> and <u>numbers</u> of <u>microorganisms</u> they contain.
2) The amounts of different <u>nutrients</u> they contain.
3) The food <u>additives</u> they contain.

This information is needed to ensure that foods are <u>safe</u>, <u>legal</u> and are <u>labelled correctly</u>.

2) Quality Assurance

Some food scientists are involved in <u>quality assurance</u> — checking all aspects of production, including:

1) <u>Ingredients</u>.
2) <u>Methods</u> and <u>equipment</u> used in food <u>manufacture</u> and <u>production</u>.
3) The <u>final product</u>.
4) <u>Packaging</u> and <u>transport</u> methods.

Quality assurance work involves visiting warehouses, distribution centres and factories. It can also involve lab work to test the quality of food and its ingredients.

3) Research

Other food scientists are involved in <u>scientific research</u> to discover new things about foods.

1) For example finding new or better ways of keeping foods <u>fresher</u> for longer and improving the <u>appearance</u> of foods.
2) They may also study the effect of different foods on <u>health</u>, e.g. the effect of <u>food additives</u> on <u>children's behaviour</u>, or the benefits of <u>olive oil</u> in protecting the <u>heart</u>.

4) Creating New Food Products

Food scientists are also involved in <u>creating new food products</u> and investigating <u>new manufacturing methods</u>. They might:

1) Invent <u>recipes</u> for new foods or drinks.
2) Decide how to <u>modify foods</u>, e.g. to make them lower in fat.
3) Plan the <u>manufacture</u> of new products.

Planning Your Investigation

There are loads of different things you could look at — but whatever you choose, the format is the same.

Write a Plan That Describes Every Stage of the Investigation

One of the most important parts of your investigation is the plan.

1) Your plan should be a series of well-ordered steps that describes every stage of your investigation.
2) It should be detailed enough to allow the investigation to be carried out by another person.
3) You should state the accuracy of any readings to be taken.
4) Include labelled diagrams where appropriate.

Example:

1) Place 10 cm³ of milk into each of four sample bottles. Place one sample bottle in a freezer (at -18 °C), one in a fridge (at 4 °C), one in a cool room (at 10 °C) and one in a warm room (at 20 °C).
2) Leave the bottles for 24 hours.
3) Make serial dilutions of the samples (see p.29): Take 1 cm³ of one of the samples and place it in a test tube. Dilute this with 9 cm³ of sterilised water. Then take 1 cm³ from this test tube, put it into another test tube and dilute with another 9 cm³ of sterilised water. Repeat this one more time so the original sample has been diluted three times.
4) Take 1 cm³ of the final dilution and pour it over a culture dish.
5) Heat a spreader in a Bunsen flame until it glows red (this sterilises the spreader). Allow it to cool, then spread the diluted milk sample evenly around the culture dish.
6) Put the lid on the dish, seal it using tape, and label it.
7) Repeat steps 3-5 for the remaining three samples of milk.
8) Incubate the plates for 24 hours.
9) Count the number of bacterial colonies that appear on each plate without opening it.
10) Dispose of the plates safely using an autoclave.

Your Plan Should Include a Risk Assessment

When planning your investigation you must also think about health and safety.

1) The equipment and method you use must be safe.
2) You'll need to carry out a risk assessment — don't forget the obvious, e.g. fire and sharp things.

Example (cont):

Risk	How to reduce risk
Risk of burns from Bunsen burner	Take care when using a Bunsen burner.
Contamination by harmful microorganisms	Sterilise equipment before and after use. Avoid hand-to-mouth/eye contact. Cover open wounds. Wear protective clothing. Wash your hands before and after the experiment. Use aseptic procedures. Sterilise the area if spills occur. Seal culture dishes using tape. Do not open culture dishes once sealed. Dispose of cultures safely using an autoclave.

Always wear gloves and protective clothing when working with microorganisms.

The best laid plans of mice and men...

One of the most important things to remember when writing your plan is that it must be in enough detail so that somebody else could carry out the investigation. It's also got to be safe — if you need reminding about all that health and safety stuff have a look back over Section 1.2 (pages 6-10).

Reporting and Evaluating

It's all very well <u>planning</u> your investigation, but you need to be able to <u>make sense</u> of all those numbers.

Use Your Results to Make Conclusions

At the end of your investigation you need to <u>interpret</u> your data and write a <u>conclusion</u>:

1) <u>State</u> what your <u>results show</u>.
2) <u>Describe</u> your <u>graphs or charts</u>, <u>identifying</u> any <u>patterns or trends</u>.
3) <u>Explain</u> your conclusions using <u>science</u> — the more <u>understanding</u> you show the better.

<u>Example (cont)</u>:

For more on serial dilutions and how these numbers were calculated see p.29.

Temperature (°C)	Number of bacterial colonies on plate	Approximate number of bacteria in 10 cm³ of milk
-18	2	20 000
4	4	40 000
10	9	90 000
20	19	190 000

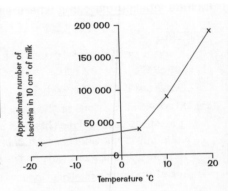

1) Milk stored at <u>higher temperatures</u> contained <u>more bacteria</u>.
2) Between 4 °C and 20 °C the graph shows a roughly <u>linear relationship</u> between the temperature and the number of bacteria — the <u>higher</u> the temperature, the <u>more</u> bacteria.
3) The results of the investigation suggest that the <u>best of the four temperatures</u> for milk storage is -18 °C. But this is not very practical for shops or consumers as the milk is frozen. For shops and consumers the results suggest that the milk should be kept as cold as possible without freezing it.
4) Milk contains <u>bacteria</u>. Bacteria contain enzymes that function at an <u>optimum temperature</u>. At low temperatures the enzymes <u>don't work</u>, or work <u>very slowly</u>. As the temperature increases, the enzymes work more <u>efficiently</u> and the bacteria are able to <u>reproduce faster</u>.

Evaluations — Describe Strengths and Weaknesses

1) <u>Describe</u> the <u>strengths</u> of your investigation.
2) <u>Explain</u> how you tried to make the test <u>fair</u>.
3) Point out any <u>weaknesses</u> in your investigation and suggest <u>ways of improving it</u>.

<u>Example (cont)</u>:

1) Using <u>serial dilutions</u> and a <u>spread plate</u> was a quick and easy method of determining the number of microorganisms.
2) The test was made <u>fair</u> by taking each sample from the <u>same source</u> and leaving each sample at its designated temperature for the <u>same length of time</u>.
3) To <u>improve</u> the results, the test could be <u>repeated</u> for each sample and an <u>average</u> taken. To improve the <u>accuracy</u> the samples stored at -18 °C and 4 °C could be <u>diluted less</u>.

Give a Vocational Application of Your Investigation

1) <u>Explain</u> how a <u>food scientist</u> might use the <u>results</u> of your investigation.
2) <u>Explain</u> why the activity is useful in the <u>workplace</u> and what <u>types of organisation</u> might use it.

<u>Example (the end)</u>:

1) The results of this investigation could be used to decide the best way to <u>store milk</u> during <u>transport</u>, in <u>warehouses</u> and on <u>shop shelves</u> — so it stays <u>fresher</u> for <u>longer</u>.
2) The procedure for testing the number of microorganisms in milk samples can be used to check that milk is <u>safe to drink</u> and complies with <u>legal requirements</u>.
3) The investigation could be carried out by <u>milk producers</u>, <u>milk retailers</u> and <u>independent food safety testers</u> to check milk quality.

Section 3.1 — Using Food Science

The Roles of a Forensic Scientist

If food science doesn't quite take your fancy, how about <u>forensic science</u>? Forensic scientists gather <u>evidence</u> that can be used in <u>courts of law</u>.

You Could Choose to Carry Out a Forensic Investigation

In your forensic investigation you should use a number of tests for comparing and matching samples. You need to include:

- a <u>plan</u> and <u>risk assessment</u> for the investigation.
- <u>conclusions</u> and <u>evaluations</u> based on your results.
- an explanation of how a <u>forensic scientist</u> might use the <u>results</u> to indicate the probability of a suspect being linked to a crime.

Forensic Scientists Carry Out a Range of Investigations

Forensic scientists work closely with the <u>police</u> and are sometimes required to go to a <u>crime scene</u> to search for <u>evidence</u> and help determine the likely <u>sequence of events</u>. Examples of <u>laboratory investigations</u> carried out by forensic scientists include:

1) Identifying and Comparing Textile Fibres

1) <u>Textile fibres</u> are strands of <u>material</u> that come from <u>fabrics</u> (e.g. clothes).
2) They are examined using <u>microscopes</u> to determine things like the <u>fibre type</u> (e.g. cotton or polyester), <u>fibre colour</u>, <u>cross-sectional shape</u> and the <u>width of the fibre</u>.
3) Textile fibres found at <u>crime scenes</u> are compared to the fibres of a suspect's clothing to see if they match.

2) Identifying and Comparing Plant and Animal Materials

1) <u>Plant</u> and <u>animal materials</u> (e.g. pollen, seeds, blood, saliva and hair) are often found at <u>crime scenes</u>.
2) Forensic scientists <u>identify</u> and <u>compare</u> these materials — the information is sometimes useful in <u>linking suspects</u> with <u>crime scenes</u> or <u>victims</u>.
3) <u>Hair</u> and <u>bodily fluids</u> like blood can be tested using <u>DNA profiling</u> — each person's DNA is <u>unique</u> so it can provide <u>very strong evidence</u> in criminal cases.

3) Analysing Blood and Urine Samples

1) Forensic scientists analyse <u>blood</u> and <u>urine</u> samples using equipment like <u>mass spectrometers</u>.
2) They can measure the level of <u>alcohol</u>, and test for the presence of <u>illegal drugs</u>.

4) Analysing Chemical Substances

<u>Chemical substances</u> found at crime scenes can be <u>identified</u> and <u>compared</u> to chemicals found on a <u>suspect's clothing or footwear</u>. Examples include:

1) <u>Firearms residues</u> (e.g. lead and nitrates).
2) <u>Perfumes</u> or <u>aftershaves</u>.
3) <u>Chemical weapons</u>.

5) Examining Paint and Glass Fragments

<u>Paint</u> and <u>glass fragments</u> found at crime scenes can provide useful evidence if they are found to <u>match</u> fragments on a <u>suspect's clothing or footwear</u>.

Planning Your Investigation

For your _forensic investigation_ you'll be given a _scenario_ — it'll be your job to find out which _suspect_ is most _likely_ to have committed the _crime_. You'll need to perform a _range_ of forensic tests on the _evidence_.

Describe the Purpose of Your Investigation

You should start your forensic science investigation by describing its _overall purpose_ — for example, 'the purpose of this investigation is to indicate the _probability_ of a _suspect being linked to a crime_.'

1) _Based on the evidence_, you need to decide what it is you want to _find out_ and what _tests_ you can use to provide this information.

2) Your investigation will involve several _different tests_ — you need to describe the purpose of _each one_.

Test	Description	Purpose
1	Comparing fingerprints found at the crime scene to those of the suspect.	To see if the suspect can be placed at the _crime scene_.
2	Comparing fibres found at the crime scene to the suspect's clothing.	To see if the suspect can be linked with the _victim_ or _crime scene_.
3	Analysing marks made by tools used to enter the crime scene.	To show if a particular _tool_ was used in the crime.
4	Comparing the ink used to print a threatening letter to the ink in the suspect's printer.	To determine whether the suspect's _printer_ could have printed the note.

Write a Step-by-Step Plan for Each Procedure

1) Your investigation will involve a number of _procedures_ — you need a _step-by-step_ method for each one.

2) You should describe _how_ the samples for the tests are _collected_ and _prepared_.

3) Your methods should be in _enough detail_ that _someone else_ could carry out the investigation.

4) You should include _labelled diagrams_ where appropriate.

Example: Method for Test 4 — Comparison of inks using chromatography

1) Mark a _pencil line_ 2 cm from the _bottom_ of a strip of _chromatography paper_.

2) Apply a dot of each sample of ink to the line, using a clean _capillary tube_ for each sample.

3) Add some _water_ to a beaker and hang the paper in the beaker so that the bottom is in the water but the ink is not submerged.

4) Leave until the solvent has _travelled up_ the paper to make a _chromatogram_ of the ink.

Your Plan Should Include a Risk Assessment

When writing a _risk assessment_ you need to:

1) List the _risks_ — all of the (likely) things that _could go wrong_.

2) List the ways you can _reduce each risk_.

Example (cont): Risk assessment for Test 2 — Using a microscope to compare fibres

Risk	How to reduce risk
Glass slides and cover slips breaking and causing cuts.	Take _care_ handling glass slides and cover slips. Clear up _broken glass_ immediately and dispose of _safely_.
Sharp equipment (e.g. mounting needle) causing injury.	Take _care_ handling _sharp equipment_.
Mountant and/or stain causing illness.	Handle _mountant_ and _stain_ using _disposable gloves_. _Wash hands_ after use. Avoid _hand-to-mouth_ / _hand-to-eye_ contact during experiment.

Remember you need to write a risk assessment for _all_ of the tests in your investigation.

Presenting Your Evidence

After you've done your investigation, you need to present your evidence and conclusions in a report.

Describe the Results of Your Tests

You should begin this section of your report by recapping all the details of the scenario, like the date and time of the crime, type of crime, etc. Then list all the evidence collected — this could include:

1) Objects from the crime scene.
2) Photographs or drawings of the crime scene
3) Fingerprints taken at the crime scene.
4) Evidence collected from suspects, e.g. fingerprints, clothing fibres.
5) Experimental evidence e.g. chromatograms.

Photograph of the crowbar found at the suspect's home.

You might want to include comments about each piece of evidence — for example the quality of the evidence, the date it was collected, and how it was packaged or obtained.

You should also explain what you found in each of your tests — but stick to the facts, e.g:

Example (cont): Results for Test 4

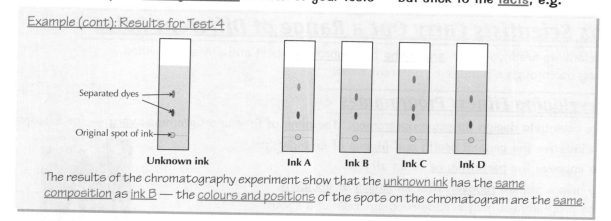

The results of the chromatography experiment show that the unknown ink has the same composition as ink B — the colours and positions of the spots on the chromatogram are the same.

Explain Your Results in the Conclusion

Your conclusion should explain what all of your results mean in the context of the case:

1) Use facts to back up your conclusions.
2) Mention any evidence that doesn't fit.
3) Explain and justify your conclusions.
4) State whether your findings support or don't support any allegations.

Example (cont):

1) The fingerprint, green fibre and chromatogram link suspect B to the crime scene.

2) The tool mark at the crime scene could have been made by a crowbar belonging to suspect B.

3) Although the shoe print evidence did not provide a match, I feel there is sufficient evidence to link suspect B to the crime scene. My findings strongly support the allegations made against suspect B.

Evaluations — Comment on the Reliability of Your Results

1) Comment on the reliability of each of your results.
2) Think about the quality of the samples and the accuracy of the procedure.
3) Try to explain any evidence that doesn't fit.

Example (end): Evaluation for Test 1

1) The fingerprint found at the crime scene was complete and clear, so I could make a confident match with the print of suspect B.

2) Fingerprints are unique for every person so it is unlikely, although possible, that the fingerprints have been incorrectly matched.

The Roles of a Sports Scientist

Love sport? Love science? Then why not investigate sports science...

There are Two Areas of Sports Science You Could Investigate

There are two different areas of sports science you could choose to investigate:

1) Devise, apply, monitor and evaluate a <u>personal fitness plan</u> for a particular sport or purpose. <u>OR</u>

2) Investigate the appropriateness of <u>materials</u> that could be used in sport for a particular purpose.

Whichever you choose you'll need to:

- describe the <u>purpose</u> of the investigation.
- describe how the investigation is <u>connected</u> with a particular sport.
- include a <u>plan</u> and <u>risk assessment</u> for the investigation.
- <u>conclude</u> and <u>evaluate</u> the investigation.
- explain how a <u>sports scientist</u> might use the <u>results</u> of the investigation.

Sports Scientists Carry Out a Range of Different Roles

Sports scientists are involved in <u>promoting</u> and <u>improving</u> sport and fitness activities, which they do through a number of different roles:

1) Developing Fitness Programmes

Sports scientists <u>design fitness programmes</u>. The <u>aims</u> of fitness programmes <u>vary</u> — for example:

1) To improve the general <u>health</u> and <u>fitness</u> of an individual.
2) To improve the <u>performance</u> of an athlete.
3) To help a patient <u>recover</u> after an <u>illness</u> or <u>accident</u>.

Fitness programmes may include recommendations about:

1) The <u>amount</u> and <u>types</u> of exercise to do.
2) How to <u>exercise effectively</u> — e.g. using <u>breathing techniques</u>.
3) <u>Diet and nutrition</u>.

2) Monitoring Changes in the Body During and After Exercise

Sports scientists use equipment such as <u>heart rate monitors</u> and <u>oxygen analysers</u> to study changes in the body <u>during</u> and <u>after exercise</u>. They can use this information to measure <u>fitness levels</u>, and to provide <u>advice</u> on <u>maintaining fitness</u> or <u>improving performance</u>.

3) Developing Diets to Improve Fitness

Sports scientists design <u>diets</u> to <u>improve health</u>, <u>fitness</u> and <u>performance</u>.
The right diet will help <u>prepare</u> the body for <u>exercise</u>, and can <u>improve performance</u> and <u>endurance</u>.

4) Developing Sports Equipment to Improve Performance

Sports scientists study the use of <u>equipment</u> and <u>clothing</u> in sports. They investigate the <u>physical properties</u> of different materials, and the <u>forces</u> that act on them, so they can pick the most suitable material for a job. Properties they investigate include things like:

1) <u>Density</u>
2) <u>Strength</u>
3) <u>Flexibility</u>
4) <u>Grip</u>

They are also involved in <u>designing</u> and <u>testing</u> new materials — such as 'smart' fabrics that help regulate your body temperature.

Devising a Fitness Plan

You can't just launch straight into it — you need to put quite a lot of <u>thought</u> into your <u>fitness plan</u>.

There are Six Things to Consider When Devising a Fitness Plan

Sports physiologists follow a number of <u>steps</u> when devising a fitness plan, these are also things that you should <u>bear in mind</u> when writing your <u>own</u> plan.

1) Decide on the Purpose

When devising a fitness plan the <u>first step</u> is to decide on a <u>purpose</u>.
For example a sports physiologist might write a plan to:

1) Improve the <u>performance</u> of a <u>weightlifter</u>.
2) Improve the <u>endurance</u> of a <u>long-distance swimmer</u>.
3) Improve the <u>health</u> and <u>fitness</u> of a non-athlete.

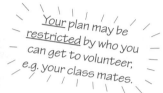
Your plan may be restricted by who you can get to volunteer, e.g. your class mates.

2) Consider the Needs of the Individual

<u>Before</u> you write your plan, the <u>circumstances</u> and <u>needs</u> of the subject need to be taken into account:

1) Consider things like their <u>height</u>, <u>weight</u>, <u>age</u>, <u>gender</u>, <u>body shape</u> and any <u>individual needs</u> before choosing <u>suitable activities</u>.
2) Think about their <u>personal interests</u> and <u>goals</u> — choose activities they <u>enjoy</u> and are <u>motivated</u> to do.

3) Choose Activities that are Suitable for the Purpose

Where appropriate the plan should include a mixture of <u>aerobic</u>, <u>anaerobic</u> and <u>flexibility</u> exercises — though the <u>proportions</u> of each type of exercise will depend on the <u>purpose</u>.

1) <u>Aerobic exercises</u> increase <u>cardiovascular endurance</u> — essential in <u>long-distance</u> running or cycling.
2) <u>Anaerobic exercises</u> increase <u>short-term muscle strength</u> — essential in weightlifting and sprinting.
3) <u>Flexibility exercises</u> improve the range of <u>muscle movement</u> — essential in gymnastics and dance.

4) Decide on the Frequency, Intensity and Duration

For each type of <u>activity</u> or <u>exercise</u> the plan should also set out:

1) The <u>frequency</u> — how <u>often</u> to do the exercise.
2) The <u>intensity</u> — how <u>difficult</u> the exercise is.
3) The <u>duration</u> — the <u>length of time</u> to do the exercise for.

Include a <u>variety</u> of activities and exercises in your plan to keep it <u>interesting</u> and to work different <u>muscles</u>.

Exercise and rest need to be <u>balanced</u>. Training every day will <u>exhaust the body</u>, while training once a month won't increase fitness. Plan exercises for <u>every other day</u>, or <u>3-4 times a week</u>.

5) Set Short-, Medium- and Long-Term Objectives

<u>Objectives</u> of a fitness plan need to be <u>clear</u> and <u>achievable</u>. Examples include:

To lift a weight by a certain date, e.g. 50 kg by the end of the month.

To run a distance in a certain time, e.g. 800 m in under 4 minutes.

To lower body fat to a certain level, e.g. to reduce body fat level by 5%.

6) Think About Health and Safety

Check that the plan is <u>safe</u>. A <u>risk assessment</u> should be carried out for all the activities:

1) <u>List the risks</u> — the things that could go wrong.
2) List the ways that <u>each risk</u> could be <u>reduced</u>.

Evaluating a Fitness Plan

Once you've devised your plan you need to <u>monitor</u> and then <u>evaluate</u> its suitability for your chosen sport or purpose — that way you can tell if it was <u>any good or not</u>.

Decide How to Monitor Your Fitness Plan to Determine Progress

You need to think of tests that'll <u>measure</u> how much <u>progress</u> is made towards the fitness plan's <u>goals</u>.

Examples of monitoring methods include:

1) <u>Observation of performance</u> — e.g. you could the record the maximum number of repetitions a weightlifter can do with a certain weight.

2) <u>Measuring body functions</u> — e.g. you could measure heart rate or breathing rate before, during and after exercise.

3) <u>Measuring body changes</u> — e.g. you could measure body weight or body fat.

4) <u>Questionnaires</u> — e.g. you could devise a questionnaire to assess well-being.

Once you have decided what monitoring methods to use you'll also need to:

1) Decide what <u>data</u> to record (e.g. resting pulse rate).

2) Decide <u>how often</u> to do the test (e.g. weekly, daily).

3) Design a suitable method to <u>record</u> the data in, e.g. a <u>table</u>.

4) Decide how to <u>present</u> your data (e.g. in graphs or charts).

Evaluate Your Plan's Suitability for Purpose

Once the plan's been carried out you need to <u>evaluate</u> it to determine how <u>suitable</u> it was.

1) Describe how <u>easy</u> the plan was — also include <u>problems</u> you came across.

2) Look back at the <u>objectives</u> you set and comment on how well they were <u>achieved</u> — if any objectives <u>weren't met</u>, explain <u>why</u> you think that was.

3) Think about how <u>suitable</u> the activities and exercises you chose were — describe any activities that were particularly <u>suitable</u> or <u>unsuitable</u>.

I'd value this E somewhere in the region of 250-300k

4) Analyse the <u>data</u> to find out if the participant <u>benefited</u>, and how — give any figures that show <u>improvement</u> (e.g. if a runner got faster). You should also include any results which <u>don't</u> show improvement and try to <u>explain</u> these.

5) Suggest <u>improvements</u> to the <u>plan</u> where possible — e.g. better ways of exercising, different activities, or changes to how often certain activities are done.

6) Suggest <u>improvements</u> to the <u>monitoring methods</u> — e.g. collecting different data, or using more accurate ways of collecting it.

Describe the Vocational Relevance of Your Plan

1) Describe <u>why</u> a sports scientist would use a <u>fitness plan like yours</u>.

2) Explain <u>how</u> a sports scientist would use the <u>results</u> (e.g. to check progress, and to improve the plan).

3) If you can, list some <u>organisations</u> that might use these activities.

My Idea of a fitness plan — running to the pie shop...

You might not like the idea of picking holes in your own work, but if you want to get those <u>top marks</u> then you have to <u>evaluate</u> both the <u>strengths</u> and <u>weaknesses</u> of your investigation, suggesting any <u>improvements</u> you'd make if you had the chance to do the investigation again. Oh how I love pie.

Planning a Materials Investigation

If you like <u>sports science</u> but you don't fancy devising a fitness plan, it's OK — there's <u>another option</u>.

Describe the <u>Purpose of</u> Your <u>Investigation</u>

You could do an investigation to find the most <u>suitable material</u> for use in a <u>particular sport</u>. If you do, you should start your investigation by <u>deciding</u> and <u>describing</u> its <u>purpose</u>, e.g.

1) Investigating density — to find the best material to make a surfboard.
2) Investigating tensile strength — to find the best material to make climbing ropes.
3) Investigating water resistance — to find the best material to make a hiking coat.

Describe how the investigation is <u>connected</u> to a <u>particular sport</u> and how the results could be used to <u>improve</u> sporting performance.

> <u>Example:</u>
>
> The density of surfboards affects their <u>speed</u> — if I can find a suitably <u>strong</u>, <u>low-density</u> material, a <u>faster</u> surfboard could be made. I will test the density and strength of a range of different woods.

You Need to Write a Step-by-Step Plan...

Your plan should describe <u>every detail</u> of your investigation. It should be in enough detail that someone else could carry it out.

1) Your plan should be a <u>series</u> of <u>well-ordered steps</u>.
2) You should include the <u>precision</u> of the <u>readings</u> you're going to take.
3) Include <u>labelled diagrams</u> where <u>appropriate</u>.

> <u>Example (cont) — Measuring the Densities of the Woods:</u>
>
> 1) Select square blocks of a range of <u>suitable sample woods</u> (square off any irregularities if necessary). Weigh each sample cube on a <u>top pan balance</u> and record the mass to 0.01 g.
> 2) Measure the <u>dimensions</u> of each block using a mm rule.
> 3) Calculate the <u>volume of the test sample</u>.
> 4) Use the formula Density (g/cm^3) = Mass (g) / Volume (cm^3) to calculate the density of each sample.

... And Produce a Risk Assessment

You'll need to do a <u>risk assessment</u> for your investigation:

1) <u>List the risks</u> — the things that could go wrong.
2) List the ways you can <u>reduce each risk</u>.

> <u>Example (cont) — Risk Assessment for Measuring the Strength of the Woods:</u>
>
> | Risk | How to reduce risk |
> |------|--------------------|
> | Wood splintering and causing cuts. | Wear goggles to protect your eyes from splinters. Clear up splintered wood immediately and safely. |
> | Weights falling and causing injury. | Take care handling heavy weights. |

Density measurements — they're also known as exams...

<u>Risk assessments</u> might not be the most interesting thing in the world but they could just <u>save your life</u>. It reminds me of the time I was carrying out an experiment when it all started to go wrong. Luckily I'd read through pages 6-10 of this very book. Remember — always expect the unexpected.

Evaluating a Materials Investigation

No investigation would be complete without <u>conclusions</u> and <u>evaluations</u>...

Use Your Results to Make Conclusions

1) Record your results in a suitable <u>table</u> and present them using <u>graphs</u> or <u>charts</u>.
2) State simply what your results <u>show</u>, identifying any <u>trends</u> or <u>patterns</u>.
3) Explain what your investigation has <u>revealed</u>.

Example (cont):

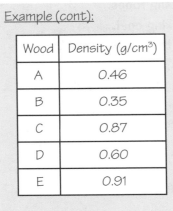

Wood	Density (g/cm³)
A	0.46
B	0.35
C	0.87
D	0.60
E	0.91

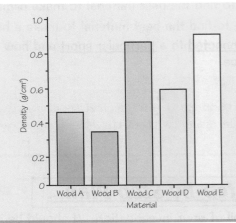

1) The results show that <u>Wood B</u> was the <u>lowest density</u> material tested.
2) <u>Woods C and E</u> were the <u>densest</u> materials tested.
3) This part of the investigation has shown that Wood B is likely to make the <u>lightest</u> and <u>fastest</u> surfboard.

Give the Strengths and Weaknesses in Your Evaluation

1) Describe the <u>strengths</u> of your investigation.
2) Explain how you tried to make the <u>test fair</u>.
3) Point out any <u>weaknesses</u> of your investigation.
4) Suggest ways of <u>improving</u> your investigation.

Example (cont):

1) Using the mm rule was a <u>quick</u> and <u>easy</u> method of determining the <u>volume</u> of the materials.
2) The tests were made fair by testing each sample in <u>exactly the same way</u>.
3) The main <u>weakness</u> was that only <u>one</u> sample of each material was tested.
4) To <u>improve</u> the results, the test could be <u>repeated</u> on two or three samples of the same material and an <u>average</u> taken.
5) To help decide if Wood B really is the <u>best material</u> to make a surfboard out of, <u>other properties</u> should be <u>measured</u> and <u>compared</u> — such as <u>resistance to corrosion</u>, <u>flexibility</u> and <u>water resistance</u>.

Explain How Your Findings Could be Used and Applied

The final part of your investigation is to give a <u>vocational application</u> — how your investigation would be used in the <u>real world</u>.

1) Explain how a sports scientist might use your <u>results</u>.
2) Explain why the activity is <u>useful</u> in the <u>workplace</u>.
3) Describe what types of <u>organisation</u> might carry out a similar investigation.

Example (end):

This method could be used by sports equipment <u>designers</u> and <u>manufacturers</u> when <u>deciding</u> on the best <u>material</u> to make a piece of sports equipment that needs to be <u>strong and lightweight</u>, e.g. a surfboard.

Tips on Producing Your Portfolios

Even if you think this stuff is <u>blindingly</u> obvious, <u>READ IT</u> anyway — humour me.
It's a list of the stuff you <u>must</u> remember when you're putting your portfolios together...

You'll Need a Portfolio for <u>Each Coursework Unit</u>

1) You'll have to produce <u>two separate portfolios</u> — one each for <u>Unit 1</u> and <u>Unit 3</u>.

2) For each unit, you'll have to write various <u>reports</u> (see pages 5, 11 and 87-97 for specific advice on the reports).

3) The portfolios are marked by your <u>teacher</u> and moderated by AQA.

4) The portfolios make up <u>60%</u> of your final grade, so they're pretty important...

I'm not impressed

Your Portfolios Should be <u>Neat</u> and <u>Easy to Follow</u>

If you hand in a <u>jumbled</u>, <u>illegible mess</u> and call it a portfolio, your teacher will <u>NOT</u> be impressed.

1) Your portfolios should be <u>well organised</u>, <u>well structured</u> and <u>tailored</u> to the tasks (so no random notes from lessons, no unidentified graphs or diagrams, no pictures of Elvis).

2) If you've got access to a computer, <u>word process</u> your reports — they're much <u>neater</u> that way, and it's easier to <u>edit</u> your work if you change your mind about something.

3) Make life easy for your marker — break up your report with <u>headings</u> to make it easier to follow.

4) If you're including any <u>graphs</u>, <u>diagrams</u> or <u>photos</u>, make sure they're clearly <u>labelled</u>.

5) There's no right or wrong <u>length</u> for a report. But they should be only as long as they <u>need to be</u> to cover everything. Don't <u>pad them out</u> for the sake of it — no one likes wading through waffle.

6) <u>Read through</u> your work carefully before handing it in (run a <u>spellcheck</u> if you're using a computer).

Make Sure It's <u>All Your Own Work</u>

Make sure there's nobody else's work in with yours. <u>I</u> know you're honest, but AQA take a very dim view of two candidates' work being <u>too similar</u>.

It's fine to include bits in your reports that come from <u>books</u> or <u>websites</u>, but you need to <u>reference</u> them — say where they come from. Your references can go at the <u>end</u> of the report.

You also need to work as <u>independently</u> as possible. The more <u>help</u> you need from your <u>teacher</u>, the lower your mark. But, saying that, it's better to do something with help than just miss it out altogether.

And Then for a Few <u>Finishing Touches</u>

Clear presentation makes your portfolio <u>easier to follow</u>... which makes life easier for the person <u>marking</u> it... which puts them in a <u>good mood</u>... which has got to be good. Here are a <u>few tricks</u>:

1) Make a <u>front cover</u> for your portfolio. It should have <u>your name</u>, the <u>course name</u> and the <u>unit number and title</u>. (There's an official cover sheet to go in front of this as well — ask your teacher.)

2) Separate the different reports with <u>header pages</u> — nothing fancy, just put the name of the report.

3) <u>Number</u> your pages. Call the first header page "page 1", then just <u>number through</u> to the end.

4) Include a <u>contents page</u> with page numbers.

5) Hole-punch everything and put it in a <u>ring binder</u>... and you're done. Woohoo!

Index

Index

Index/Answers

Answers

Revision Summary for Section 2.2 (page 30)

23) If six colonies grew from one tenth of the final dilution, there were approximately:
60 bacteria in the final dilution,
600 bacteria in the second dilution,
6000 bacteria in the first dilution.

So 1 cm³ of the milkshake contained approximately **6000 bacteria**.

Revision Summary for Section 2.4 (page 41)

16) Plaster of Paris.

19) E.g. size of hand, length of fingers, number of fingers, presence of rings or scars, lines on the palm.

25) If you have no suspect to match a fingerprint to, the print can be compared with the national fingerprint database to try and find a match from known criminals.

Bottom of page 43

a) MgO

b) Li_2O

c) Na_2SO_4

Revision Summary for Section 2.5 (page 50)

5) a) FeO

b) Fe_2O_3

c) $CaCl_2$

d) Na_2CO_3

17) a) barium sulfate + iron(II) chloride

b) silver chloride + zinc nitrate

21) a) The blue precipitate means it contains copper. The white precipitate means it contains sulfate ions.
The powder is copper sulfate.

b) $CuSO_4$

Revision Summary for Section 2.8 (page 77)

8) $1.3 \times 24 \times 75 = 2340$ Kcal/day

10) professional runner, builder, waitress, secretary

13) $82 \div 1.77^2 = 26$: overweight — this might not be an accurate description because Barry is an athlete, which means he'll probably have a lot of muscle, which weighs more than fat.

Revision Summary for Section 2.9 (page 85)

24) a) Polythene (because it's hard and strong to protect the cyclist, but lightweight so as not to slow the cyclist down).

b) Leather (because it's tough, strong and durable enough to withstand being dragged across rough ground).

c) Carbon fibre (because it's very light for it's strength, and for a high-performance piece of equipment you're willing to pay for it).